RARE & WONDERFUL

Treasures from Oxford University Museum of Natural History

Kate Diston and Zoë Simmons

First published in 2018 by the Bodleian Library
in association with Oxford University Museum of Natural History
Broad Street, Oxford OX1 3BG
www.bodleianshop.co.uk

ISBN 978 1 85124 484 3

Thanks are due to the following for permission to reproduce images in this book:

27 © Bodleian Libraries, University of Oxford, Douce T subt. 15 (1)
158 © NRF-SAIAB. Sketch by Marjorie Courtney Latimer.
Illustration provided by the South African Institute for Aquatic Biodiversity
160–61 Clockwise from top left © The Tolkien Trust 1995; © The Tolkien Trust 1978; © The Tolkien Estate Ltd 1937. Reproduced by kind permission
167 Photographed with permission of the Estate of Anthony Stones
168–9 Photographed with permission of The Nobel Foundation
179 © Chong Chen
189 © University of Warwick
190–91 © Arthur Anker
194–5 © Bastian Reijnen
207 © Mike Peckett

Special thanks to the Bodleian Imaging Studio, especially John Barrett and Nick Emm, and Katherine Child for their expert photography of many of the specimens. Thanks also to Scott Billings for his stunning photography of the Museum and to Sarah Joomun for her photograph of the Lyell Collection.

Cover design by Dot Little at the Bodleian Library
Designed and typeset in 11 on 16 Caslon by illuminati, Grosmont
Printed and bound through Great Wall Printing Co. Ltd, Hong Kong, on 157 gsm FSC® certified Neo Matt paper

British Library Catalogue in Publishing Data
A CIP record of this publication is available from the British Library

Acknowledgements

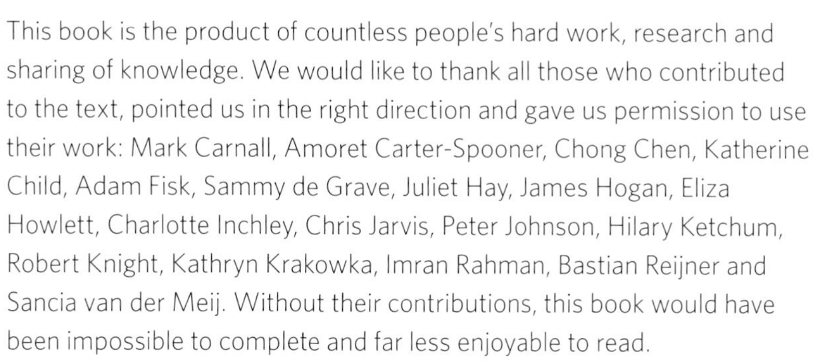

This book is the product of countless people's hard work, research and sharing of knowledge. We would like to thank all those who contributed to the text, pointed us in the right direction and gave us permission to use their work: Mark Carnall, Amoret Carter-Spooner, Chong Chen, Katherine Child, Adam Fisk, Sammy de Grave, Juliet Hay, James Hogan, Eliza Howlett, Charlotte Inchley, Chris Jarvis, Peter Johnson, Hilary Ketchum, Robert Knight, Kathryn Krakowka, Imran Rahman, Bastian Reijner and Sancia van der Meij. Without their contributions, this book would have been impossible to complete and far less enjoyable to read.

We would also like to thank Professor Paul Smith, Director of Oxford University Museum of Natural History, and Wendy Shepherd, Head of Operations, for their support in producing this book.

Many thanks to the Bodleian Library Publishing team for their patience and guidance, in particular Janet Phillips, Leanda Shrimpton and Susie Foster. Special thanks go to Samuel Fanous for seeing the potential of the museum's collections and for supporting this book from conception to print.

Finally, we would like to acknowledge all the people who have looked after the collections in this museum since its construction, and all those that will do so in the future. Your care of the collections will ensure that they are always available for everyone to enjoy, learn from and use. This book would not have been possible without that care.

Introduction

Since the opening of the University Museum in 1860 (later Oxford University Museum of Natural History), a spirit of scientific discovery, provocative thought and innovative learning has been at the heart of all its activities. As one of the earliest purpose-built science museums in the world, it is a unique space devoted to the collection and preservation of objects from the natural world, and to teaching about them. This mandate continues over a century and a half later, driving all areas of work in the museum, from education and research to collecting and public engagement.

At the time the museum was first conceived in the 1840s the University of Oxford's scientific collections were primarily housed in the old Ashmolean Museum, located in the current Museum of the History of Science on Broad Street. They included the Tradescant Collection, containing the famous dodo specimen, as well as other assemblages compiled by contemporary readers and lecturers, such as the renowned geologist William Buckland and entomologist Frederick William Hope, whose names will appear many times in the following pages.

In addition to the old Ashmolean Museum, there were also notable collections of scientific material in Christ Church and the Bodleian Library, with smaller collections in numerous other colleges and departments. These were primarily stored as 'cabinets of curiosities', as was common in the period. This was a time of significant discoveries of new species as well as a developing understanding of the natural world. The shifting focus of academics

OPPOSITE *The University Museum*, 1860. Engraving by J.H. Le Keux.

OVERLEAF The museum under construction in the 1850s.

to the study of the new sciences, coupled with the lack of care the specimens were receiving as part of these dispersed collections, led to an appeal to Oxford University to erect a new building dedicated to the sciences. This new museum would both provide suitable storage for its collections and establish the centre of teaching and research for science at Oxford.

The petition to build the museum had been championed by anatomy professor Henry Wentworth Acland, who felt strongly that all Oxford students should have the opportunity to be educated in the sciences. After a similar appeal had been rejected in 1847, the university finally voted in favour of the new museum in 1849. In 1854, Oxford University committed £40,000 to the construction of Oxford University Museum and a contest was held to design the new home for the sciences at Oxford. Thirty-two entries were submitted for the architectural design, and the winner was young Irish architect Benjamin Woodward, who had based his design on the façades of medieval Belgian cloth halls. Largely influenced by the design principles of John Ruskin, decorative features were an important, but costly, element of the design.

This was a critical period, not just at Oxford, but across Europe, in the development of science as an academic discipline. New departments were forming and the scientific collections of universities were growing. When the museum officially opened in 1860 it housed all the science departments in Oxford at the time: Astronomy, Geometry, Experimental Philosophy, Mineralogy, Chemistry, Geology, Zoology, Anatomy, Physiology and Medicine. The names of these departments are still painted over the office doors on the lower gallery, though they are now home to more modern museum functions such as Operations, Public Engagement, IT and the Director. It is difficult to believe that some of the most famous and influential science departments in the world began in

a single room, but the museum swiftly became the centre for science at Oxford. It did not take long for science to outgrow this relatively small building, however, and the fledgling departments moved out, over time, to new purpose-built accommodation around the rapidly developing science area, leaving behind the collections but retaining links between the museum and the new science departments.

While the science departments no longer call the building home, the role of the museum is still positioned at the centre of science at Oxford. Public engagement has become increasingly important within the world of research and university museums are uniquely placed to communicate science in interesting and meaningful ways. While this may be a newly developing trend across the sector, the museum has a long history of being a place to discuss and debate cutting-edge science. One of the most important events in the history of the museum happened around the same time as its official opening. On 30 June 1860, the British Association for the Advancement of Science held its annual meeting at the museum. The topic on everyone's mind was a discussion surrounding the recent publication of Charles Darwin's *On the Origin of Species*. Though no one recorded the exact discussion, legend grew about the heated exchange between Samuel Wilberforce, Bishop of Oxford, and Thomas Huxley, a London biologist known as 'Darwin's Bulldog'. This exchange would later become known as the 'Great Debate' and marked the shift in thinking from a predominantly religious to a scientific view of the natural world. Evolution was one of the most fiercely contested theorems and set the precedent for the museum's commitment to disseminating newly developing scientific ideas.

Eminent collectors, whose names will appear again and again throughout this text, included figures such as Charles Darwin, William Buckland, Mary Anning, William John Burchell, Henry Acland, John Obadiah Westwood and Frederick William Hope. They and many others played a formative role in natural sciences as both subjects in their own right, and in how collections of specimens came together, moving between researchers and being dispersed to various museums and institutions around the country. For better or worse their legacies have shaped how we view nature and the environment, with our perceptions of material culture being inextricably bound up with their personalities, such was the strength of their impact at the time. There are others, however, whose stories have perhaps become a little lost, such as that of Katherine Ethel Pearce, photographer and dipterist, but are, nonetheless, just as important. Wherever possible these have been teased out from various archives and collections to reveal some of the hidden beauties and wonders of the museum's collections.

Founder Henry Wentworth Acland and John Ruskin in 1893. Photograph by Sarah Angelina Acland.

Researchers of the natural world are not the only people to have been inspired by the museum or its collections. Writers and artists have also been captivated and intrigued. The dodo, in particular, seems to have captured people's imaginations, inspiring authors new and old, from Jasper Fforde to Lewis Carroll. Illustrations and artworks by local artists such as Katherine Child, Jennifer Mathison and Kelley Swain, former poet in residence, continue to be produced and exhibited. 'The Dodo Gavotte', taking its lead from 'The Lobster Quadrille' written by Lewis Carroll, was composed especially for the publication of this book (see p. 13), in order to celebrate the diversity of creativity that is stimulated in the wider world through contact with the museum and its collections.

While the ethos of the museum has not changed, much else has. The way the collections are stored has improved vastly, with the use of modern materials for storage and the monitoring of environmental conditions to ensure the collections last for future generations. New technologies are used to record and store information about the collections, making that information accessible to more people across the globe. The museum now welcomes more than 700,000 visitors through its main doors each year; countless more than would ever have been imagined when it first opened. Visitors are not only growing in numbers but are also increasingly diverse, with local families and tourists joining academics and Oxford undergraduates. The museum is not just a home to Oxford University's science collections, it is a centre for inspiring everyone to learn more about natural history and promote a new generation of important scientific discovery.

What the museum collects has also expanded and changed focus. It now holds over seven million scientific specimens – though every last beetle hasn't been counted! Overall, five million of those are entomological specimens, or insects, as well as approximately half

PREVIOUS SPREAD The interior of Oxford University Museum of Natural History today.

a million palaeontological (fossil) specimens and half a million zoological specimens. The museum also holds an extensive collection of archival material, both related to the specimen collections and to important historical naturalists such as William Smith, William Jones and James Charles Dale, and their work. This now amounts to millions more objects than the original founding collection and has, over time, filled most of the building and numerous off-site stores. Collecting has always remained a key activity for the museum and continues to this day, ensuring that specimens and archives accurately represent our natural world today, and that they will provide information for the science of the future.

Natural history museums are important and irreplaceable repositories of the natural world, both living and extinct. Most visitors are not aware of the enormous trove of specimens stored behind the scenes, with only a very small percentage of the collections on display at any given time. While the average visitor does not get to see these, staff, researchers, artists and volunteers are always at work behind the scenes. Much can be discovered about our environment and the creatures that exist within it by studying a natural history museum's specimens and the information recorded about them. Taxonomic studies can provide a greater understanding of the relationships between living organisms and the evolution of life. Ecological studies can help understanding of patterns in species distribution over time. All these would be impossible without natural history collections such as the one held in Oxford.

Museums are often seen as static places with unchanging collections, but natural history museums in particular are ever-changing and growing. The image of an independently wealthy nineteenth-century gentleman naturalist in the field, notebook and

nets in hand, often comes to mind when we think of collecting natural history specimens. While this romantic notion no longer resembles the reality of modern science of the natural world, scientists are still actively collecting. Many find it hard to believe, but newspaper headlines still regularly announce the discovery of new species. The types, or first described specimen of these species, are often deposited in natural history museum collections, remaining as an important record of the discovery and as a sample against which all further biological finds of the species are

Decorative artwork on the iconic roof of the museum.

compared. The species known to scientists and stored in museums today represent only a small proportion of life on Earth and, as new life is discovered, it is collected, recorded and made available for future scientists to access.

Selected from over seven million specimens, and several thousand documents, objects and works of art, the items described in this book represent only an extraordinarily small percentage of the museum's collections. The selection presented here does not represent the most valuable, most important or best known, though many of these are included. Instead, they all tell unique stories about natural history, of the history of science, of the people involved in it both new and old, and of Oxford University Museum of Natural History itself. They reflect the importance of collecting, preserving and making natural history collections available to researchers and the public, and demonstrate what kind of information can be gleaned, both from the past and for the future, about the natural world that we all live in and of which we are all a part. Beginning with the museum's door, welcoming visitors in, and ending with the famous glass roof, securely containing all within it, the items in this book will take you on a journey across the collection, through time and around the world.

OVERLEAF *The dodo and other birds,* nineteenth-century copy after Roelant Savery, 1626.

DoDo & Given
by G. EDWARDS
F. R

The Dodo Gavotte

'Will you visit my Museum?'
said the Dodo to the Bear.
'There are great things to discover
if you spend some hours there.'

'Beneath these vaulted arches
all of Science is displayed:
with grand Victorian drama,
"Truth to Nature" is our aim.'

Will you, won't you, will you
Love Geology?

'Within this light-drenched Cabinet
of plants in iron and stone
you can see the largest Sperm Whale jaw
and the first-named-Dino bones.'

'Where "Darwin's Bulldog" and
"Soapy Sam"
sparred, with elocution,
whilst Lady Brewster fainted
at the case for evolution.'

Will you, won't you, will you
Love Zoology?

'At the old cliffs of Lyme Regis
Mary found an Ichthyosaur:
it now lies in my Museum
with a thousand fossils more.'

'You can cuddle a tarantula,
feel stones become pyrite,
wonder at the Thylacine,
see ancient Trilobites.'

Will you, won't you, will you
Love Palaeontology?

'For mapping complex molecules
in two and three dimensions
Dorothy made history –
won Nobel Prize attention.'

'Come visit my Museum
with the gemstones and the bees,
you can study all of Science
and you'll never want to leave!'

Will you, won't you, will you
Love Entomology?

'Come to where extinction lives,
come see where fish can fly,
where insects see in infra-red:
where Wonder lights our sky.'

Will you, won't you, will you
Love Natural History?

Kelley Swain

The abandoned archway

The museum is rich in decorative features typical of Victorian Neo-Gothic design, reflecting perfectly the time and place of its construction. Many of these features were influenced and executed by some of the artists of the Pre-Raphaelite Brotherhood, as well as its supporter John Ruskin, who had strong connections to Oxford and the university. While many features, such as the stunning glass roof, were executed to design, a number were never completed due to a lack of funds (see pp. 120 and 206). The detail the architect and designers of the museum sought to include in building Ruskin and Acland's 'cathedral to science' was much more costly than originally anticipated.

One of the features that was never completed is the archway above the main doors to the museum. Two designs were proposed for this archway, the first by one of the founding members of the Pre-Raphaelite Brotherhood, Thomas Woolner, and the other by both Woolner and his contemporary, John Hungerford Pollen. Both featured religious iconography, specifically Adam and Eve. The first design was not used at all, likely because it failed to marry religion and science, and represented Adam and Eve in a traditional way, being cast out of Eden for their pursuit of knowledge. The second design, which is partially finished, also featured Eve and Adam, but with an angel at the apex, holding both a book and a living cell. It has been said that James and John O'Shea, the Irish stonemasons employed to carve this intricate feature, were dismissed over other carvings made in the porch, though we now know they left the job unfinished because the university ran out of funds (see p. 119).

There is little doubt Christian beliefs and values played an important role in the lives of many Victorian scientists, but Woolner and Pollen's design demonstrates that these two opposing ideals were equal in the minds of those that founded this museum, and that science was playing an increasing role in their search for an understanding of the natural world.

Dodo

Now an icon of extinction, the dodo, *Raphus cucullatus*, has, for many, come to symbolize much more than simply a species of flightless pigeon. First recorded in 1598 and extinct by 1662, the species is now surrounded by layers of myth and story. The overriding narrative is of a fat, waddling and somewhat stupid bird that was driven to extinction by the gastronomic efforts of visiting sailors to its island home.

Endemic to Mauritius, the dodo was a ground-dwelling bird, with no mammalian predators to disturb its nests. By all accounts its meat was distasteful, so rather than having been driven to extinction through human consumption it is likely that its demise came about from a combination of degradation of its habitat and the introduction of pigs to the island by Dutch sailors using the site as a layover on long sea journeys. Pigs are voracious ground

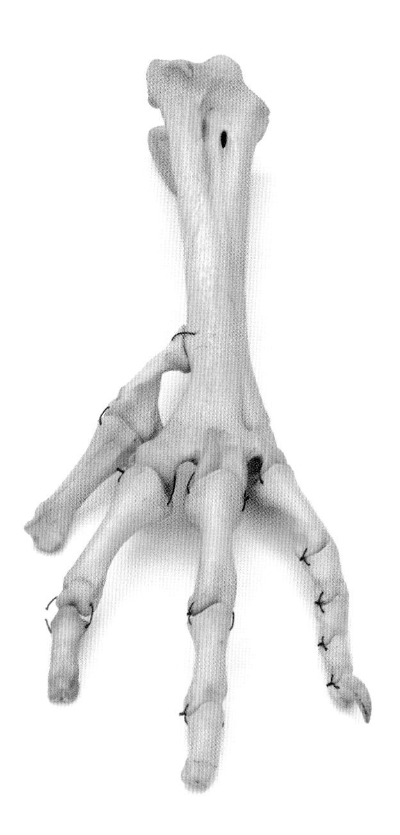

scavengers, against which the dodo would have had no defence, having evolved in isolation.

Given the dodo's popularity in modern culture it is perhaps surprising to learn that only a very few pieces of dodo bird exist in the world today. No complete skeleton remains, though composite skeletons, constructed from sub-fossil remains and casts, have been produced. The few models that have been created take their inspiration from a handful of drawings and illustrations drawn from live specimens.

Unique among the remains then, are those held by this museum. Arguably the most treasured holding, the specimen represents the only preserved soft tissue remains of an entire species. This was once part of a complete taxidermy mount which was on display in the seventeenth century in John Tradescant's London museum. The Tradescant Collection was bequeathed to Elias Ashmole, founder of the Ashmolean (see p. 29), whence the remains came to Oxford University Museum of Natural History.

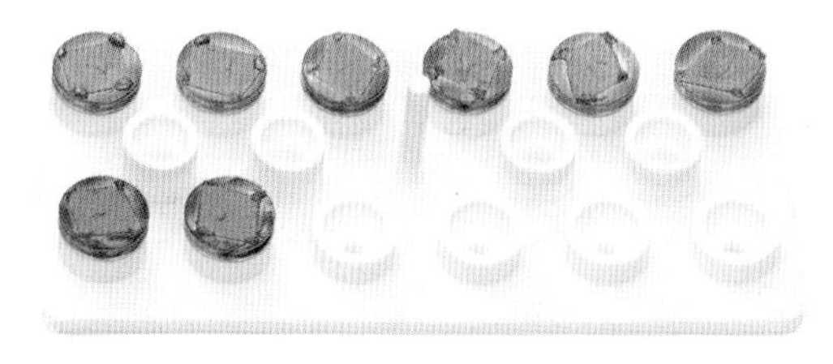

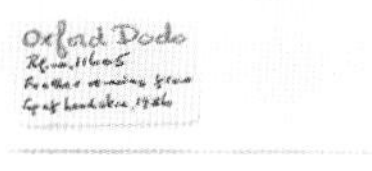

Ammonites

Ammonites are one of the most recognizable types of fossil, often being among the first fossils that schoolchildren learn to identify by their characteristic spiral shape. One might easily assume, therefore, that they are simple to understand. In fact, they belong to a group of fossils that show some of the greatest diversity in the fossil record. They are often used as index fossils, allowing geologists to identify different types of rock, or strata, from each other and across vast distances. This is only possible because of their varied and nuanced differences.

The ammonite is an extinct marine mollusc that lived between 240 million and 66 million years ago, ranging across several geological ages. Different types of ammonite can be identified by characteristic bulges and indentations and the patterns of the sutures or seams in their chambered shells, and by the siphuncle, a tube running inside the outer edge of the shell that connected all the chambers. These features also differentiate them from living nautiloids, to which they are related.

The part of the animal that we normally see today is its preserved shell. When living, this would have held a squid-like creature with long tentacles which would have been used to catch food. Inside its shell there was a chambered section or phragmocone, and the animal could move up and down in the water by controlling air pressure within the chambers via the siphuncle.

The shell is now made visible by being filled with mineral deposits of a different colour to the surrounding rock. This process of fossilization

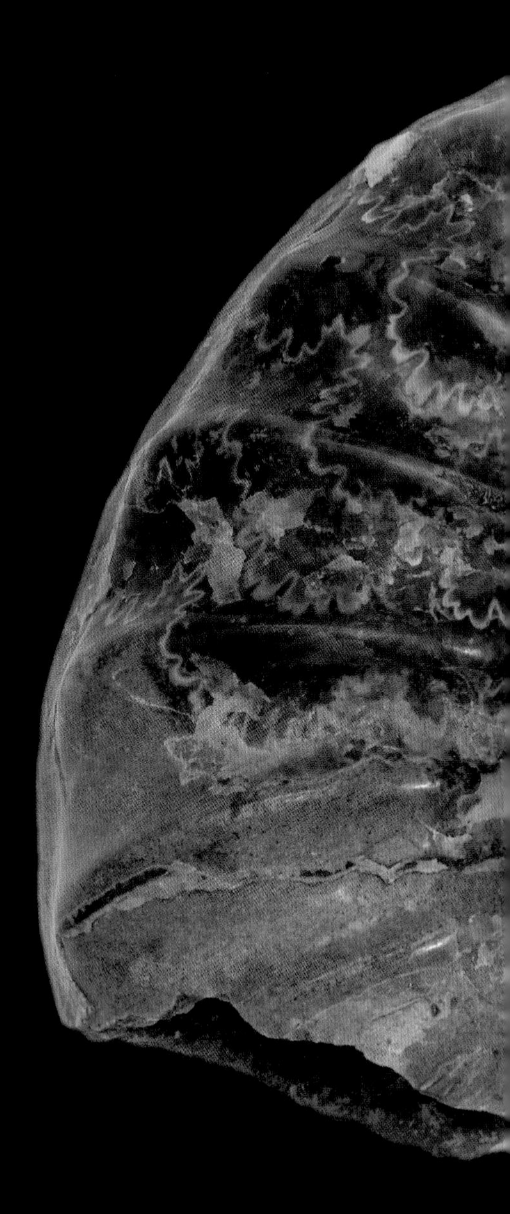

takes thousands of years, together with a series of events in specific circumstances that, over time, lead to preservation and then discovery. When an animal dies its fleshy body decomposes, leaving only bone, or shell in the case of an ammonite. If it is lucky enough not to be scavenged, or broken down by the elements, it will become covered by debris, often in a river or seabed, and over very long periods of time will be compressed. As the bone or shell dissolves, which takes much longer than flesh, it will shrink. The space left by the dissolved shell or bone will make room for other elements to move in, creating mineral deposits. These fit into the space left behind, eventually forming the fossil that can be seen today.

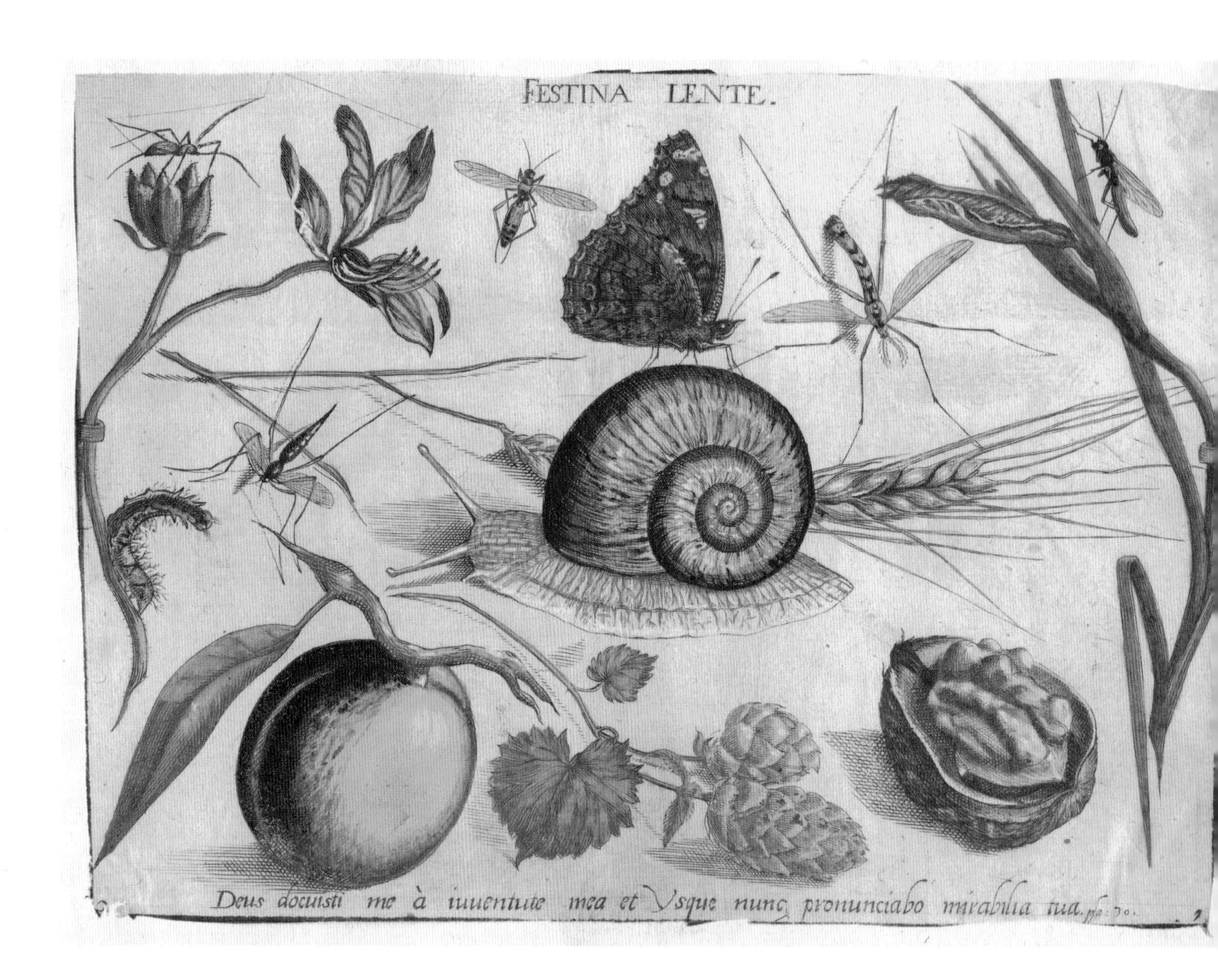
FESTINA LENTE.
Deus docuisti me à iuuentute mea et Vsque nunc pronunciabo mirabilia tua. psa: 70.

Before 'still life'

Archetypa studiaque patris Georgii Hoefnagelii is one of the oldest and most beautiful books in the museum's library collection. Published in Frankfurt in 1592 by son and father Jacob and Joris Hoefnagel, it includes forty-eight beautifully detailed and colourful engravings of plants, small animals and insects, divided into four sections, each containing twelve plates. The engravings were made by son Jacob, only nineteen at the time, and were based on paintings by his father, the famous painter and miniaturist Joris Hoefnagel. The book was intended as a religious work featuring mottos contemplating God's plan for creation and his influence on the natural world, a common theme at the time.

What is immediately striking about this book is the very lifelike nature of the representations. It is clear that they must have been completed from observations of specimens. While that is a common practice in both artistic and scientific drawing today, it was not so in the sixteenth century. It has been argued that the drawings in this book represent the foundation of what would become the still-life genre practised in the Golden Age of Dutch painting.

The book itself also has an interesting history, in addition to its content. This particular volume was rebound at some point in the twentieth century, unfortunately in a modern binding with the pages cut out and remounted. We also know its provenance: it was donated to Oxford University by Reverend Frederick William Hope, along with his enormous founding collection of entomological material, documents and books. Before it belonged to Hope it was in the library of Isaac D'Israeli, Esq., father of the Prime Minister Benjamin Disraeli. It is possible that Hope purchased the book from D'Israeli, but this is not the only connection between them. Hope married the wealthy heiress Ellen Meredith who had very recently rejected a marriage proposal from Benjamin Disraeli, stating that 'a life as the wife of a politician would have been a very dull one indeed' (correspondence in Frederick William Hope Collection).

Fantastical four-footed beasts

One of the oldest books in the museum's library is Edward Topsell's *The History of Four-Footed Beasts and Serpents*, 1658. Topsell (*c*.1572–1625) was an English clergyman whose fascination with the natural world unfortunately did not extend much further than his writing desk, and who was better known by his contemporaries for his works on religious and moral issues. He is most famous, however, for compiling the first work of natural history published in Britain in the English language, though not necessarily for its scientific accuracy.

The History of Four-Footed Beasts is an encyclopaedia of types, outlining and illustrating all 'known' beasts and creatures. It includes many mundane and ordinary animals that most Britons of the time would have readily recognized, such as cats, horses and hedgehogs. In addition to these, it also illustrated exotic species that few in Europe would ever have seen, including elephants, rhinos, lions and chameleons. Some of the most striking and unusual animals described in Topsell's book, however, are those that are mythical. The unicorn is one such fantastical creature described by Topsell, featuring the distinctive, narwhal-like tusk coming from the head of a horse, which we are advised can guard against poisons if ground and added to water. The manticore is another, less commonly known, beast, described as a creature having the head of a man with three rows of sharp, shark-like teeth, the

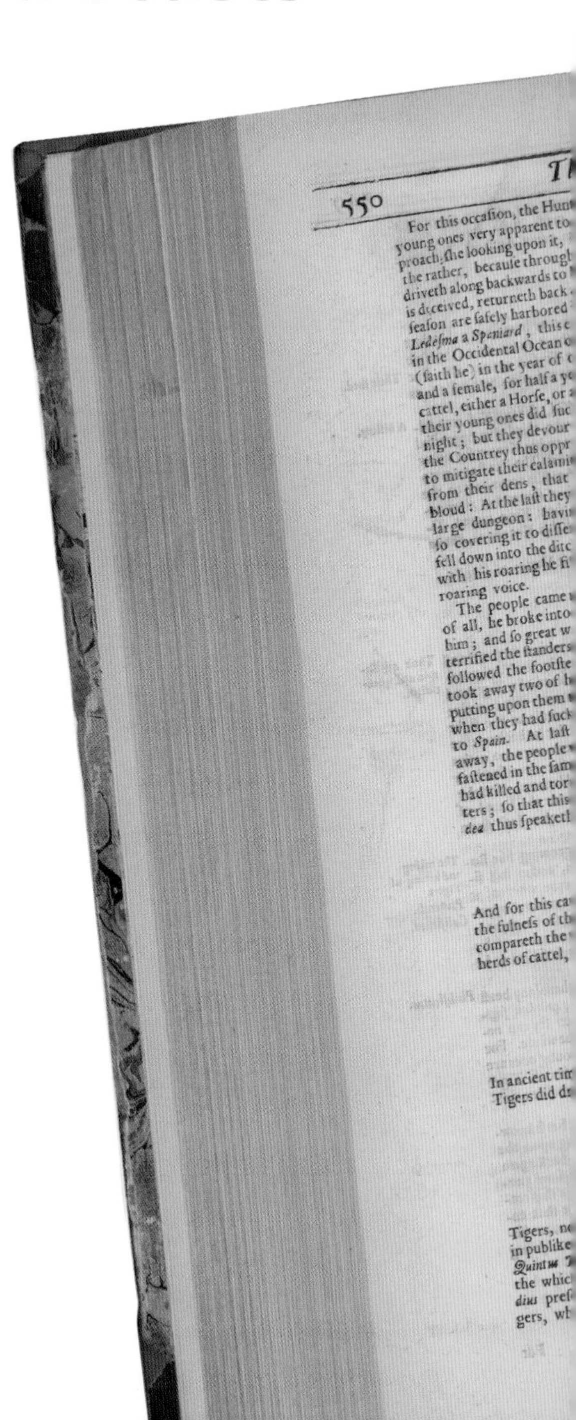

550

For this occafion, the Hun
young ones very apparent to
proach, fhe looking upon it,
the rather, becaufe through
driveth along backwards to
is deceived, returneth back
feafon are fafely harbored
Ledefma a *Spaniard*, this c
in the Occidental Ocean o
(faith he) in the year of
and a female, for half a ye
cattel, either a Horfe, or a
their young ones did fuc
night; but they devour
the Countrey thus oppr
to mitigate their calami
from their dens, that
bloud: At the laft they
large dungeon: havi
fo covering it to diffe
fell down into the ditc
with his roaring he fi
roaring voice.
The people came
of all, he broke into
him; and fo great w
terrified the ftanders
followed the footfte
took away two of h
putting upon them
when they had fuck
to *Spain*. At laft
away, the people
faftened in the fam
had killed and tor
ters; fo that this
dea thus fpeaketh

And for this ca
the fulnefs of th
compareth the
herds of cattel,

In ancient tim
Tigers did d

Tigers, n
in publike
Quintus
the whic
dius pref
gers, wh

nd sphears of glass, wherein they picture their
f these they cast down before her at her ap-
hat her young ones are inclosed therein, and
is apt to rowl and stir at every touch, this she
h it with her feet & nails and so seeing she that
r her true Whelps; whilest they in the mean
e on some shipboard. It is reported by Johannes
d female Tiger. In the Island Dariene, standing
some eight days sail from Hispaniola, it fell out
aid Island was annoyed with two Tigers, a male
e was no night free, but they lost some of their
re, or a Hog, and Swine, and in the time that
n to go abroad in the day time, much less in the
first of all meet with another beast: At length
d them to devise a remedy, and to try some means
ht out all the ways and paths of the Tigers to and
ance upon the raveners for the loss of so much
way, this they cut asunder and digged deep into a
ey strewed upon the top of it little sticks and leaves,
rneath, then came the heedless Tiger that way, and
akes, and pointed instruments as they had there set,
out, and the Mountain sounded with the eccho of his

eat and huge stones upon his back killed him, but first
the stones, Weapons, and Spears, that were cast against
was half dead, and the bloud run out of his body, he
g upon him. The male Tiger being thus killed, they
here the female was lodged, and there in her absence
ward changing their mindes, carryed them back again,
hains, and making them fast in the same den, that so
they might be with pleasure and safety conveighed in-
was come that they should be taken forth to be sent
they found neither young nor old, but their collers
them, whereby it was conceived that the envious mother
s, rather then they should fall into the hands of the hun-
ed in horrible cruelty, and for this occasion is it that Me-

tum me de Tigride natam,
opulos gestare in cor de videbor.

hout singular wit by the Poets, that such persons as satisfie
f revenge, are transformed into Tigers. The same Poet
betwixt two advantages unto a Tiger betwixt two preys or
of them to devour, in this manner;

diversa valle duorum,
e, mugitibus armentorum,
s ruat, & ruere ardet utroq;
eus dextra levave feratur.

cated to Bacchus, as all spotted beasts were, and that the said
e did hold the rains; and therefore Ovid saith thus;

ru quem summum texerat uvis,
djunctis aurea lora dabat.
ner;
ntem Bacche pater tue
es indocili jugum collo trahentes.

t mindes and untamable wildeness, have been taken, and brought
he first of all that ever brought them to *Rome*, was *Augustus*, when
s were Consuls, at the dedication of the Theater of *Marcellus*,
im out of *India*, for presents (as *Dion* writeth.) Afterwards Clau-
; and lastly *Heliogabalus* caused his chariots to be drawn with Ti-
when he said;

Picto

Picto quod juga delicata collo,
Pardus sustinet, improbæq; Tigres,
Indulgent patientiam flagello.

Ledesma of whom we spake before affirmeth, that he did eat of the Tigers flesh that was taken in the ditch in the Island *Dariene*, and that the flesh thereof was nothing inferior to the flesh of an Ox, but the *Indians* are forbidden by the laws of their Countrey, to eat any part of the Tigers flesh, except the hanches. And thus I will conclude this story of the Tiger, with the Epigram that *Martial* made of a Tiger, devouring of a Lion. (Eating of Tigers.)

Lambere securi dextram & consueta magistri,
Tibris ab Hyrcano gloria rara jugo,
Seva ferum rabido laceravit dente Leonem:
Res nova, non ullis cognita temporibus.
Ausa est tale nihil sylvis dum vixit in altis:
Postquam inter nos est, plus feritatis habet.

Of the UNICORN.

WE are now come to the history of a Beast, whereof divers people in every age of the world have made great question, because of the rare vertues thereof; therefore it behoveth us to use some diligence in comparing together the several testimonies that are spoken of this beast, for the better satisfaction of such as are now alive, and clearing of the point for them that shall be born hereafter, whether there be a Unicorn; for that is the main question to be resolved.

Now the vertues of the horn, of which we will make a particular discourse by it self, have been the occasion of this question, and that which doth give the most evident testimony unto all men that have ever seen it or used it, hath bred all the contention; and if there had not been disclosed in it any extraordinary powers and vertues, we should as easily believe that there was a Unicorn in the world, as we do believe there is an Elephant although not bred in *Europe*. To begin therefore with this discourse, by the Unicorn we do understand a peculiar beast, which hath naturally but one horn, and that a very rich one, that groweth out of the middle of the forehead, for we have shewed in other parts of the history, that there are divers beasts, that have but one horn, namely some Oxen in *India* have but one horn, and some have three, and whole hoofs. Likewise the Bulls of *Aonia*, are said to have whole hoofs and one horn, growing out of the middle of their fore-heads. (Many beasts with horns, improperly called Unicorns. *Solinus. Ælianus. Oppianus.*)

Likewise in the City *Zeila* of *Æthiopia*, there are Kine of a purple colour, as *Ludovicus Romanus* writeth, which have but one horn growing out of their heads, and that turneth up towards their backs. *Cæsar* was of opinion that the Elk had but one horn, but we have shewed the contrary. It is said that *Pericles* had a Ram with one horn, but that was bred by way of prodigy, and not naturally. *Simeon Sethi* writeth, that the Musk-cat hath also one horn growing out of the fore-head, but we have shewed already that no man is of that opinion beside himself. *Ælianus* writeth, that there be Birds in *Æthiopia* having one horn on their fore-heads, and therefore are called *Unicornes*: and *Albertus* saith, there is a fish called *Monoceros*, and hath also one horn. Now our discourse of the Unicorn is of none of these beasts, for there is not any vertue attributed to their horns,

and

body of a lion and the tail of a scorpion, which it used to paralyse its victims before devouring them whole.

Topsell largely copied the descriptions in *The History of Four-Footed Beasts and Serpents* from works written in other languages. This was common for many of the early English-language works that attempted to compile all known information on a specific

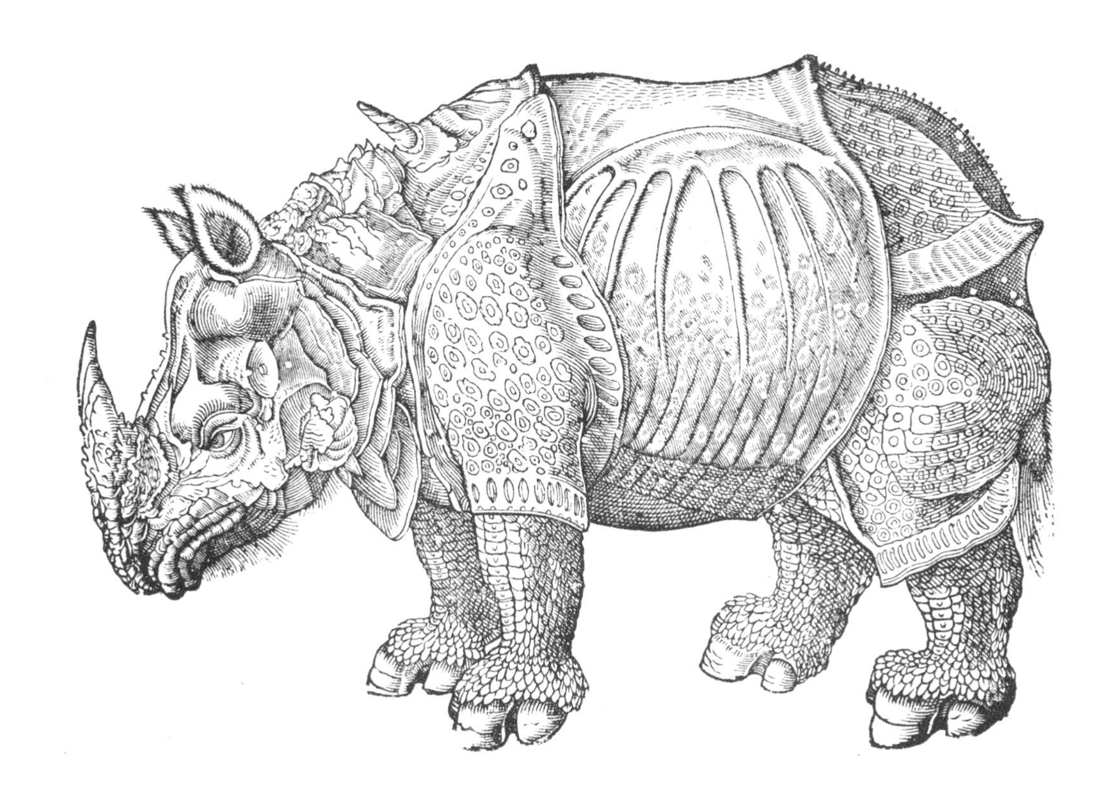

topic. In particular, much of this book is taken from the Swiss work *Historia animalium* by Conrad Gesner. Topsell's approach to writing a compilation of known knowledge without applying his own research and experience would soon be replaced by a more scientific model for describing the natural world; one that would include no encounters with manticores and unicorns.

Tradescant's boar

The 'Tradescant boar' is the oldest known warthog specimen in the world. For many years, however, it was overlooked in the collection, and it was not until a relatively recent revision of the genus that the specimen's significance was uncovered. It took long hours of painstaking research to reveal the history of this skull. The task was made more difficult through loss of early documentation, the only remaining clues to its provenance being two small labels, one a metal tag and the other pasted to the skull.

Previous to the discovery of this specimen it was assumed that the first warthog specimen had arrived in Europe around 1766 when Peter Pallas described the now extinct Cape warthog. This specimen is thought to predate that by at least eighty years.

This common warthog skull is just one of the items from the Tradescant Collection, which is the oldest collection of natural history in England and includes the only soft-tissue remains of the now famous dodo. A father-and-son team, Tradescants John senior and John junior, collected a large number of botanical, natural history and ethnographic materials from around the world, establishing what eventually became known as the Musaeum Tradescantianum. Commonly referred to as The Ark, it was one of the first collections that could, for a small price, be viewed by the general public. After the death of Tradescant senior in 1638, the collection passed to Tradescant junior, who eventually bequeathed it to Elias Ashmole, for whom the Ashmolean Museum was named.

Ashmole was a former student of Oxford University and offered to present all his collections, which, of course, included the Tradescant material, to the university on the condition that a proper museum be built to house them. This purpose-built museum was erected on Broad Street and included chemistry laboratories, rooms for undergraduate lectures and a space to display the various collection materials. It is now occupied by the Museum of the History of Science.

Robert Plot's fossils

One of the earliest publications recording the existence of fossils and attempting to classify, explain and understand them is Robert Plot's *A Natural History of Oxfordshire*. Published in 1677 and reprinted in 1705, this work presented the collection of fossils Plot had amassed during his long career as the first Professor of Chemistry at Oxford University and first Keeper of the Ashmolean Museum, the university's only museum at the time.

When Plot wrote his book, the origin of fossils was not fully understood. Most still believed that the Earth was just a few thousand years old and the theory of evolution was nearly 200 years away from first being published by Charles Darwin. Most fossils were explained using the recently discovered process of crystallization and were thought to resemble living creatures only by chance. There were some, however, that Plot recognized as being too similar to living forms to be a coincidence.

One of the most interesting examples is a section of leg bone, or femur. Without realizing it, Plot was the first to describe a dinosaur bone in English, a femur of a theropod similar to *Megalosaurus*, named 150 years later by another Oxford academic, William Buckland (see pp. 58–61). With no concept of the enormous, now long extinct, reptiles to fall back on, Plot tried to intellectualize the finding of a huge bone that clearly resembled the leg bone of many living creatures, including humans. His explanation could be considered quite fanciful by today's scientific standards, but demonstrates the influence that religion and folklore still had on academic study of that period.

> It remains, that (notwithstanding their extravagant Magnitude) they must have been the bones of Men or Women: Nor doth any thing hinder but they may have been so, provided it be clearly made out, that there have been Men and Women of proportionable Stature in all Ages of the World, down even to our own Days.

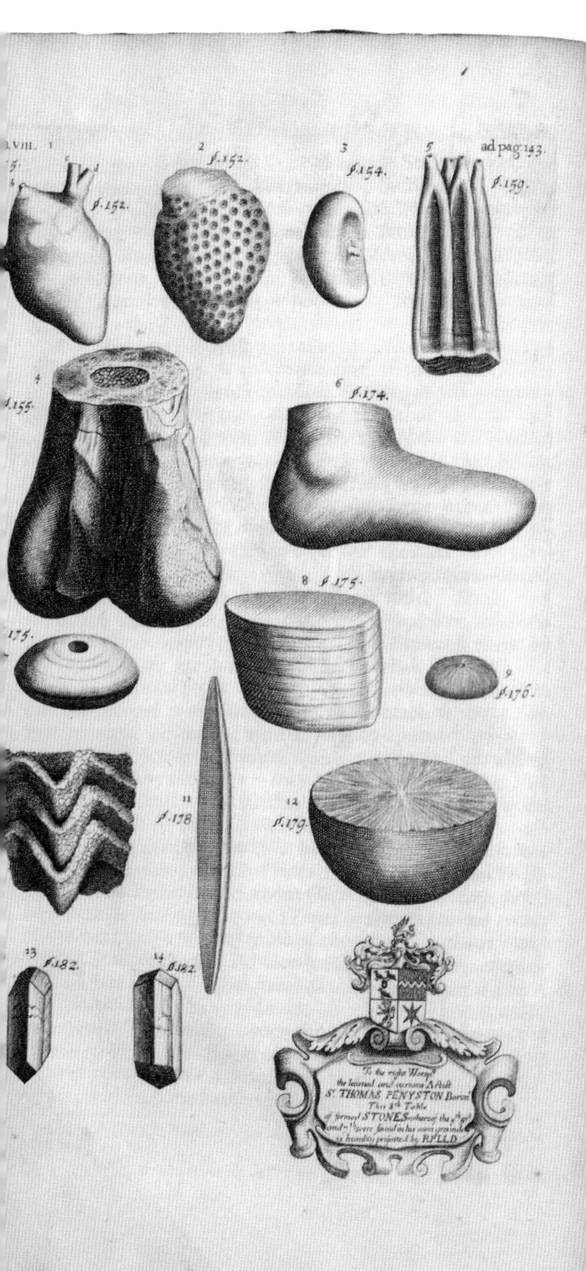

184. Only I must beg leave first to advertise the *Reader*, that what I have ascribed to Dr. *Merret* concerning the *Toad-stone*, *sect.* 148. I have found since the first Printing of that Sheet, seemingly also given to the Learned Sir *George Ent*, by the no less Learned Sir *Thomas Brown*, in his *Pseudodoxia Epidemica*[h], to whether more rightly, let them contend. And that since the first Printing the Beginning of this *Chapter*, I received from the Right Worshipful Sir *Philip Harcourt* of *Stanton-Harcourt*, two kinds of *Selenites*, though of the same Texture, yet much differently formed from any there mention'd; both of them being *Dodecaedrums*, but the *Hedræ* too as much different from one another, as from any of the former: The first sort of them being made up of two *Rhomboideal* sides, four oblong, and as many shorter *Pentagons;* and two small *Trapeziums*, one half whereof are represented *Tab.* 8. *Fig.* 13. And the second, of two oblong *Hexagons*, four oblong *Trapeziums*, four oblong *Parallelograms*, and two large *Pentagons*, one half whereof are also represented *Fig.* 14. In both which it is to be understood, that the *Hedræ* at the Ends of each *Stone*, are opposed by two others like them, not according to the Breadth, but Length of the *Stone*. The two *Pentagons* at the Top of the *Stone*, *Fig.* 13. being opposed by two others like them, behind the small *Trapezium* at the Bottom of it; and the small *Trapezium* at the Bottom, by another like it behind the two short *Pentagons* at the Top: and so the oblong *Parallelograms*, and large *Pentagons* at the Ends of the *Stone*, *Fig.* 14.

[h] Pseudodox. *Epidem. lib.* 3. *cap.* 13.

ADDITIONS *to* CHAP. V.

§. 1. A large Account of *Formed Stones* see in *Britan. Bacon.* p. 75, 76.

§. 17. The *Asteria* or *Star-stone.*] Of these *Asteriæ* see *Cambden*'s Discourse in *Lincoln-shire*, p. 536.

§. 17. In *Glocester-shire* they are taken, *&c.*] *Asteriæ* at *Belvoir-Castle* in *Leicester-shire*, and *Purton* in *Gloc. Britan. Bacon.* p. 81, 82.

§. 63. The *Turbinated* or Wreathed kind of *Stones.*] I am told that the Sands of the *Sea* somewhere in *Italy*, viewed by a *Microscope* by Dr. *Blackmore*, appeared all of this Form.

§. 66.

The Bath white butterfly

Significant in the history of entomology as the oldest surviving pinned insect specimen, this butterfly is now over 300 years old. It is a testament to the various owners and curators who have taken care of it that it is in such good condition for its age, still identifiable as *Pontia daplidice*.

Given various common names in early works, including 'Vernon's Half-Mourner' by James Petiver, the first person to assign and publish English common names for insects, this species is now known as the Bath white. This name was attached to the species by entomologists and butterfly enthusiasts after it was illustrated by William Lewin and appeared for the first time in print. The original representation whence the name came was in the form of a piece of needlework executed by a young woman from Bath after studying a specimen. Whilst the needlework may now have been lost to posterity, the name lives on.

The species is common in Southern and Central Europe and is known to migrate to northern France and beyond. It appears rarely in Britain, however, and, as such, sighting one even nowadays causes a lepidopterist's heart to flutter.

This specimen was captured by naturalist Thomas Vernon in Gamlingay near Cambridge, and its exact collection date still remains something of a mystery despite the label stating '1702'. A mention in Petiver's book *Musei Petiverion Centuria*, published in 1699, hints at it having been collected even earlier.

Now the specimen resides in the Dale Collection, and while the riddle of its exact collection date may never be fully resolved, it is safe to say it dates from at least 1702, a fact not unnoticed by the staff of the museum, who threw a 300th birthday party to celebrate in 2002.

Life cycles of insects in Suriname

Maria Sibylla Merian (1647–1717) was a remarkable entomological scientist and an incredible artist. In her most famous work, *Metamorphosis Insectorum Surinamensium*, 1705, she demonstrates both these skills to an exceptional level.

The copy held by the museum library is a folio-sized volume containing sixty beautifully hand-coloured engraved plates illustrating the life cycle of various insects found in the Dutch colony of Suriname. Merian completed the work based on notes and drawings she had made in the field, an unusual pursuit for a woman in the late seventeenth century, but Merian was very unlike the typical women of her time.

Merian came from a German family of publishers. After marriage at the age of eighteen, she chose to keep her surname and ran her own business for a number of years. She also published several artistic and scientific volumes before working on *Metamorphosis*. After leaving her husband when she was in her mid-thirties, she moved to the Netherlands and began work as a scientific illustrator in Amsterdam, drawing from specimens collected across the Dutch colonies. She saw the limitations of this method of working, however, as it would not allow for an understanding of the life of the specimen in question, and she sought to undertake fieldwork to improve her understanding.

At the age of fifty-two, Merian decided to sail to the Dutch colony of Suriname with her youngest daughter, Dorothea. It was an incredibly brave move for a woman on her own, with no official patronage. The colony had only recently been settled, and to judge from her accounts of her time there, it was a harsh environment in which to live. During her time there, however,

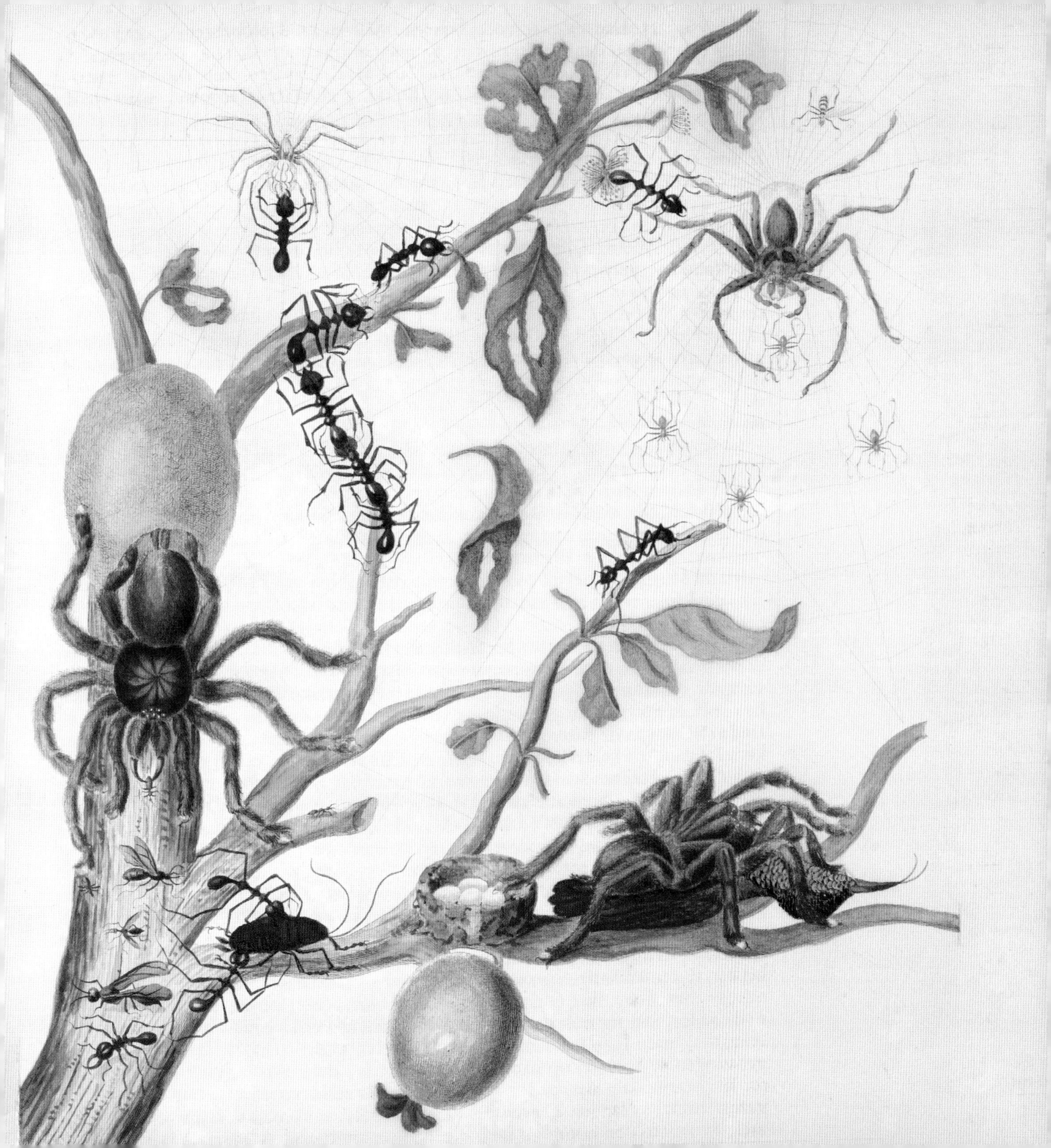

VERANDERING DER

DE I. A

DE *Ananas* zynde de voornaa
eerſte van dit werk en van m
ende vertoond, gelyk in het volg
gecoleurde bladeren dicht onder d
ken vercierd, de kleine uitſpruitzel
vrucht afgeplukt is, de lange blad
gras groen, aan de kanten wat rood
verige is de cierlykheid en fraeiheid
van de *Heeren Piſo en Markgrave* i
Deel van de Hortus Malebaricus,
ſterdamſche Hof, als ook van ander
daar mede niet ophouden, maar tot

Kakkerlakken zyn de bekendſte a
de en ongemakken, die ſy allen In
wollen, linnen, ſpys en drank,
deze vrucht zeer genegen zyn, ſy
rond geſpinſt omgeeven, als zomm
ryp zyn, en de jonge volmaakt, by
ge Kakkerlakjes met groote raſſighe
weeten ſy in kiſten en kaſten te kon
dan alles bederven, ſy worden dan
blad te zien is, van coleur bruin en
hebben, dan barſt haare huit op der
daar uit, week en wit, de huit blyft
lak was, maar leedig van binnen.

Op de andere zyde van deze vruc
draagen haar zaad onder haar lyf in
ſe het ſakje vallen, om beter te kon
getjes, en veranderen als de voorga

De bezondere benaamingen, waar meed
was van verſcheide Antheuren werd genaan
by den andere te vinden, in myn flora M.

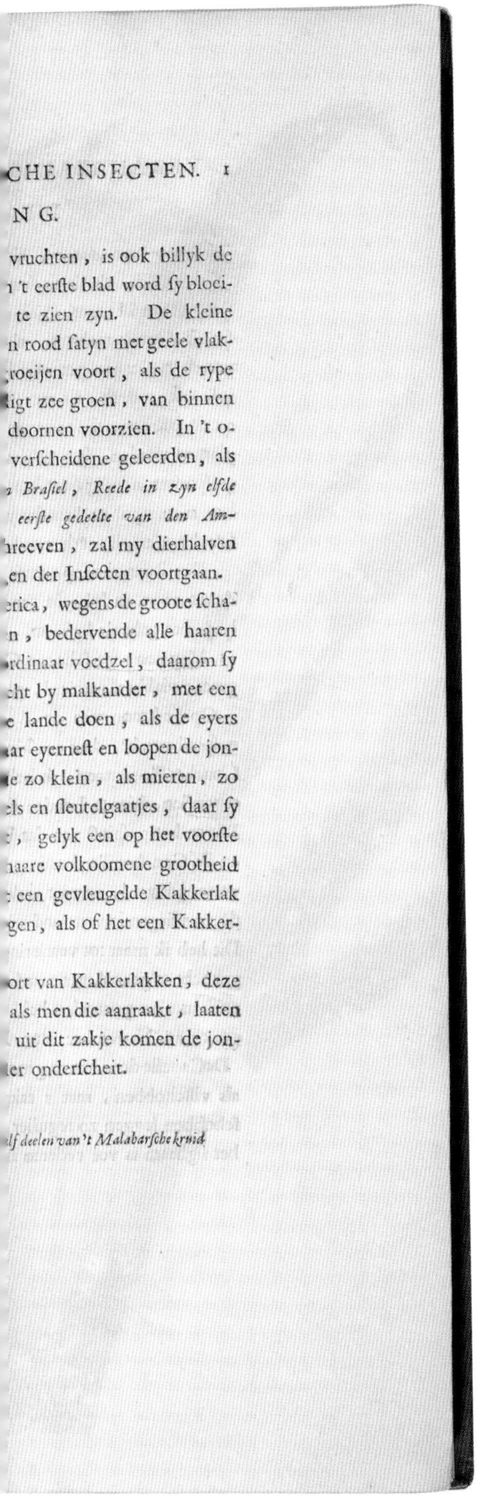
CHE INSECTEN. 1
N G.
vruchten, is ook billyk de
'n 't eerste blad word sy bloei-
te zien zyn. De kleine
n rood satyn met geele vlak-
roeijen voort, als de type
igt zee groen, van binnen
doornen voorzien. In 't o-
verscheidene geleerden, als
Brasiel, Reede in zyn elfde
eerste gedeelte van den Am-
hreeven, zal my dierhalven
en der Insecten voortgaan.
erica, wegens de groote scha-
n, bedervende alle haaren
rdinaar voedzel, daarom sy
cht by malkander, met een
e lande doen, als de eyers
ar eyernest en loopen de jon-
e zo klein, als mieren, zo
els en sleutelgaatjes, daar sy
, gelyk een op het voorste
aare volkoomene grootheid
een gevleugelde Kakkerlak
gen, als of het een Kakker-
ort van Kakkerlakken, deze
als men die aanraakt, laaten
uit dit zakje komen de jon-
er onderscheit.
lf deelen van 't Malabarsche kruid

she was able to make and record some important scientific observations, and once she returned to Amsterdam, three years later, she began work on her seminal publication.

In each of Merian's illustrations we see a carefully laid-out scene depicting the life cycles or metamorphoses of various insects and spiders, from egg through larva to pupa and finally adult. They also feature plants or other insects that would typically be a source of food to the species, as well as common habitats. These were some of the first ever recorded observations of this type.

The Works of the Lord are Great, Sought out of all them that have Pleasure therein, Ps CXI. v. 2.

The celebrated Aurelian

One of the fashionable pursuits of the late eighteenth and early nineteenth centuries for both gentleman and ladies was that of entomology, or the study of insects. In particular, this meant collecting and discussing specimens, with, generally, a focus on moths and butterflies. It had not always been so, however, and many early enthusiasts were criticized for their unusual pastime, which required them to head outdoors wearing strange garb and carrying bags full of collecting equipment. Their unusual behaviour is well depicted in the frontispiece of Moses Harris's *The Aurelian or Natural History of English Insects*, with a young man, likely Harris, posing in the field.

The number of enthusiasts grew considerably over time, with many societies founded and disbanded in Europe, including over ten in London alone in the decades leading up to the nineteenth century. One in particular, the Aurelian Society, was one of the oldest zoological societies in the world, founded in the 1720s. It met regularly in a London pub, the Swan Tavern in Exchange Alley. Its members collected and documented insects, specifically butterflies and moths, and discussed both the specimens and the art of collecting. Very little is known about the society as the Swan Tavern, which housed its collection, documentation and papers, was destroyed in a fire less than thirty years later. The group never reformed, perhaps disheartened by the loss.

What we do know about the society comes from just a few publications. Once such work is by the nephew of a member, Moses Harris, only a child at the time the group existed. Harris would also author the most complete work on British butterflies and moths (Lepidoptera) of the time, published in

1766, *The Aurelian or Natural History of English Insects*. He was also among the group which resurrected the Aurelian Society in 1762, as he described in his book: 'phoenix-like our present society arose out of Ashes of the Old'. This society would last for an even shorter time, but it was the first of many entomological societies that would be founded in the years that followed.

Harris collected from the field, and the knowledge he gained from this work was evident in his publication, though he did not name any new species in this particular book. He often included observed features or traits in his descriptions, such as the flight patterns of specific species. His illustrations were very accurate in their depiction of the specimens, though they were stylized in their composition, often containing unrelated objects and insects and botanic specimens. They were, and remain, very aesthetically pleasing. Though this book was not innovative or revolutionary in content, it would become the most celebrated work on Lepidoptera of its time, and remains among the great classical entomological publications in British history, no doubt because of its charm and beauty.

PL.XLI.
h
b
q
f
g
k
i
m
e
r
s
o
l
d
p
n
c
a
M. Harris Fecit Oct. 21. 1763
To Mr. Andrew Peter Dupont.
This plate is Humbly Dedicated by his most Obedient Servt. Moses Harris.

Out of this world

Meteorites are among the few types of material from outside our atmosphere that can be found on Earth. They are in fact rocks that originate from an object, such as a comet or planetary body in outer space, which has fallen to Earth. In order to be found on this planet, they must survive passage though the Earth's atmosphere and the enormous impact when hitting the ground. While they are relatively rare geological finds, many end up in museum collections around the world. Oxford University Museum of Natural History

holds a small but impressive collection of both meteorite 'falls' and 'finds', so-called according to whether they were observed falling or found after the fact.

The Krasnojarsk, or 'Pallas Iron', meteorite was the first to be accepted by the scientific community as having origins from outside Earth or its atmosphere. The date of fall is unknown, but it was recorded in 1772 by German naturalist Peter Pallas in Siberia. Its enormous size, weighing over 700 kilograms, and its appearance, led him to the conclusion that it must have come from space, as it was too large to have been moved and resembled no rock type in the region. This notion was widely believed in folklore at the time, but had never been proven. The museum holds a 1,700 g mass of this meteorite. It is not known how the museum acquired this, but it is likely from the Richard Simmons Collection.

The Krasnojarsk meteorite came from the asteroid belt, a band of planetary debris orbiting around the Sun between Mars and Jupiter. It was formed from the same raw materials from which the Earth is made, and at about four and a half billion years old, is the same age as the Earth itself. Meteorites like this one shed light on the deep composition and early history of the Earth.

The Nakhla meteorite fell near the village of Nakhla in the Abu Hommos district, Egypt, on 28 June 1911. It contains minerals that formed in water, with isotopes that point to a very different origin in the solar system – that is, Mars. It belongs to a group of Martian meteorites known as 'Nakhlites'. It also has a thin black fusion crust, a characteristic feature of meteorites, formed as the heat of friction melted the surface during the meteorite's journey through the Earth's atmosphere. This 52 gram piece of the Nakhla meteorite was presented to the museum by the Egyptian government, and is among the meteorites regularly used to teach Oxford's undergraduate students.

Fabricius E.S 150 Archippus

Alis repandis fulvis venis margineque albo punctato nigris: anticis apicis fulvis. —

habitat in America

Cram: P. 3 *Erippus*
206 *Plexippus*

Jones' Icones

At first glance, *Jones' Icones* looks like any other of the thousands of modern folios lining the bookshelves of Oxford. This unassuming, six-volume set, rebound countless times in its 250-year history is, however, one of the museum's greatest treasures. This original manuscript, containing over 1,500 beautiful paintings of butterflies and moths (Lepidoptera), is virtually unknown to history, but is one of the most important documents in the early study of insects.

William Jones of Chelsea (*c.*1745–1818), a wealthy retired wine merchant from London, completed his *Icones* as a work of leisure over a thirty-year period in the late eighteenth century. Well connected among the eminent naturalists of his day – including his close friend Sir James Edward Smith, founder of the Linnaean Society – Jones was given easy access to the growing numbers of Lepidoptera specimens being brought back from the colonies and sold to collectors. He travelled only in London, visiting the homes of collectors and recording their impressive collections of butterflies and moths in the form of exquisitely detailed gouache and ink paintings. He meticulously recorded each

Fabricius N° 22 Remus *Cramer Tab. 10*

Alis dentatis subconcoloribus nigris, posticis utrinque maculis flavis marginalibus. — habitat in Amboinâ

collector, and his depictions of their specimens would become a lasting record of the earliest Lepidoptera collections in Britain.

Jones' Icones is also of great importance to the history of taxonomy, or the science of classifying the living world. Johan Christian Fabricius, a student of Linnaeus, named over 10,000 insects according to his teacher's system of classification, and on hearing of Jones' manuscript, he travelled to his home to see this impressive work. Upon examining the paintings Fabricius encountered over 200 species that were unknown to science, and named these using the Linnaean system. This is an unusual practice in science, as species are usually named from a specimen rather than a depiction of one. These rare identifications are known as iconotypes. Fabricius would publish these names in his work *Systematica Entomologica* in 1775, thereby making these identifications official in scientific terms.

Work continues on this remarkable manuscript, with attempts being made to identify all the butterflies and moths depicted; to determine if the original specimens depicted still survive; to confirm their status as iconotypes; and to identify them by their modern species names.

The mysterious Mr Jones

One of the greatest treasures in the museum's archival collections was created by one of the collections' least-known figures, William Jones of Chelsea. While very little is known of the man who created *Jones' Icones*, the beautiful six-volume manuscript outlining some of the most important butterfly and moth collections in Britain at the end of the eighteenth century, we do know that he corresponded with influential natural scientists in London at the time, including Sir James Edward Smith, founder of the Linnaean Society.

The Linnaean Society was founded in 1788, deriving its name from the creator of the system for naming species systematically using binomial nomenclature, Carolus Linnaeus. It was established to further and promote the study of natural history and remains the oldest existing natural history society in the world. We know that William Jones of Chelsea was one of the first members of the society because a receipt for his membership survives in the archives of his collection.

Genoa July 7th 1787.

Dear Sir

Perhaps you may wonder at my not having sooner answered your last favor; perhaps too you may have done me the honor to be a little displeased, or jealous if you please, an honor I value more than you think; but the fact is I arrived at Milan 6 weeks later than I expected, & coming soon after to Genoa, I was attacked with a pleurisy, which although soon removed, made me unable to write much. I am now quite well, & in a weeks time shall go to Turin & from thence thro' Switzerland to Paris, where (by the bye) I shall rejoice to find a letter from you the end of Augt. chez Mons.r Broussonet N.o 57 rüe des blancs manteaux. — Am glad I have satisfied you for the present about the Linn.n Society; so we need say no more about that matter till we meet, when you shall give your assistance to that project of mine in any manner or degree you please; at least I rely on your counsel.

I have said I am proud to find myself capable of exciting your jealousy, & my reason for saying so is that I am no stranger to that feeling myself, but there are very few people that I honor with it. Indeed I rather wish never to feel it again, for it is in me connected with such a degree of esteem & affection as scarcely

There is also a letter to Jones, dated 7 July 1787, from James Smith, discussing his idea for a learned society devoted to the furtherance of natural history, just months before he would found the Linnaean Society. Clearly Jones must have been someone Smith was keen to discuss this with and Smith invited him to express his interest in joining such a group.

While we know so little of Jones today, his connection to influential and well-known scientific figures leaves us with hope that new information, archival records and even collections may emerge that shed more light on this remarkable early lepidopterist.

Saved by a beetle

Pierre André Latreille (1762–1833) was an eminent French zoologist who was described by Fabricius as 'the foremost entomologist of his time'. At the age of sixteen Latreille moved to Paris to study, where he became interested in natural history, learning from great scientists including René Just Haüy and Jean-Baptiste Lamarck and collecting insects wherever he went. Despite this devotion to the natural world, he started studying to become a priest and left the Grand Séminaire of Limoges as a deacon in 1786.

In 1790, during the French Revolution, the French government introduced an obligatory civic oath for priests, requiring them to pledge allegiance to France or face being deported. For unknown reasons Latreille missed the deadline for swearing the oath and was condemned in 1793. He spent two years in prison, first in Brive and then in Bordeaux, but was saved from deportation thanks to a small beetle.

A doctor visiting the prison was surprised to see Latreille crawling on the floor looking at an insect. Latreille explained that he was examining *Necrobia ruficollis*, a rare and important beetle. Clearly impressed by his knowledge, the doctor sent the beetle to a young naturalist named Jean Baptiste Bory de Saint-Vincent. Bory de Saint-Vincent was already aware of Latreille's entomological work and sought help to secure his release from prison. Latreille expressed his gratitude towards those who saved his life that day in his 1806 book *Genera crustaceorum et insectorum*:

> This is an insect very dear to me, for in those unhappy times when France groaned under the weight of calamities of every kind, I was much indebted to the kind assistance of Bory de St Vincent and Dargelas, but chiefly to this insect for my liberty and salvation.

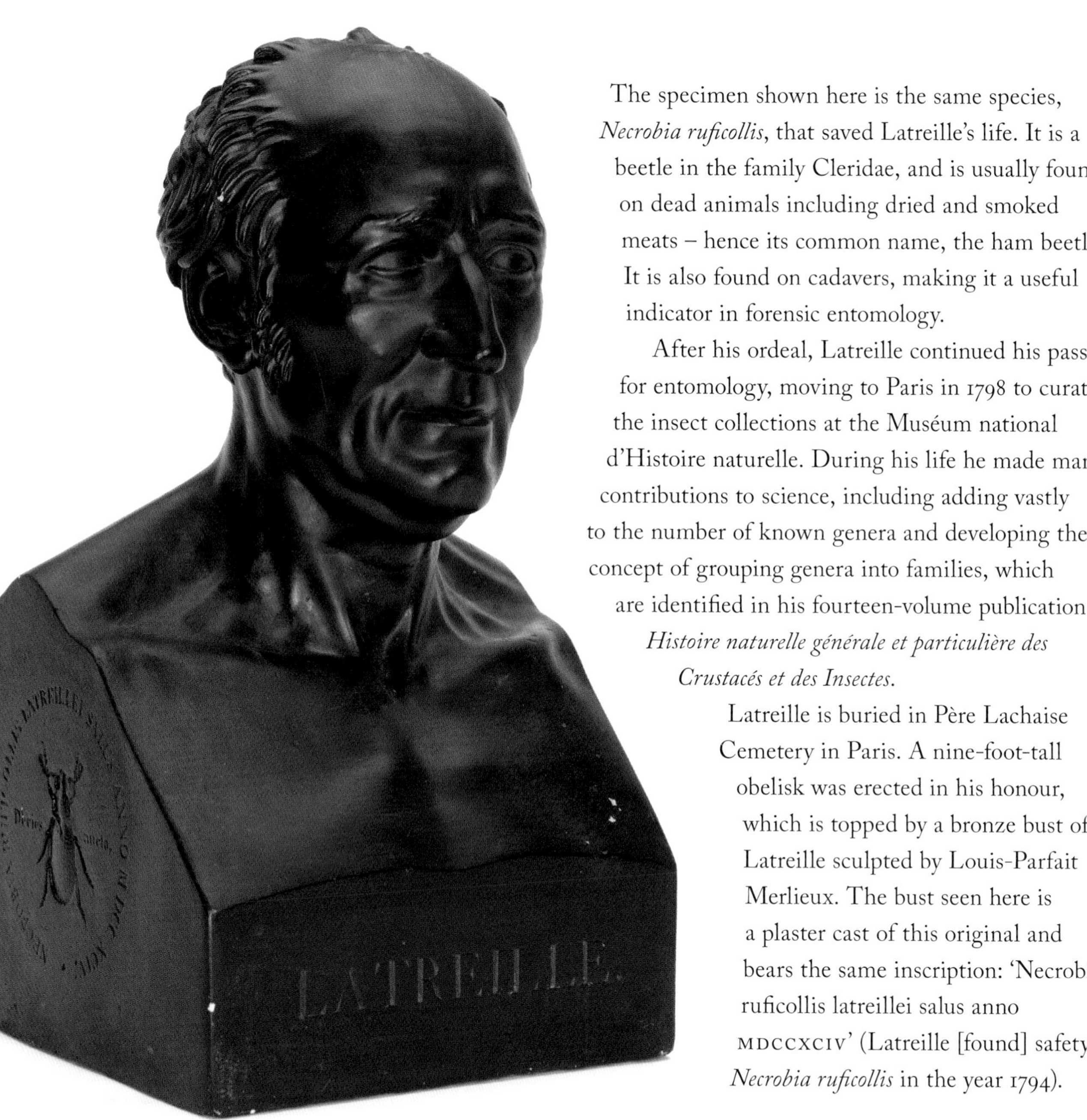

The specimen shown here is the same species, *Necrobia ruficollis*, that saved Latreille's life. It is a beetle in the family Cleridae, and is usually found on dead animals including dried and smoked meats – hence its common name, the ham beetle. It is also found on cadavers, making it a useful indicator in forensic entomology.

After his ordeal, Latreille continued his passion for entomology, moving to Paris in 1798 to curate the insect collections at the Muséum national d'Histoire naturelle. During his life he made many contributions to science, including adding vastly to the number of known genera and developing the concept of grouping genera into families, which are identified in his fourteen-volume publication *Histoire naturelle générale et particulière des Crustacés et des Insectes.*

Latreille is buried in Père Lachaise Cemetery in Paris. A nine-foot-tall obelisk was erected in his honour, which is topped by a bronze bust of Latreille sculpted by Louis-Parfait Merlieux. The bust seen here is a plaster cast of this original and bears the same inscription: 'Necrobia ruficollis latreillei salus anno MDCCXCIV' (Latreille [found] safety in *Necrobia ruficollis* in the year 1794).

The Giant Irish elk

Megaloceros giganteus was the largest deer that ever lived. It was nearly twice the size of its nearest living relative, the fallow deer (*Dama dama*), standing over two metres tall at the shoulders and with antlers up to three and a half metres from tip to tip. The fossil record shows that the male *Megaloceros giganteus* shed its enormous antlers each year. Their impressive size would have allowed the males to intimidate rivals and studies have suggested that, given their functional morphology and musculature, they also would have been able to use them to fight. The female Giant Irish elk did not have antlers, and this could account for their scarcity in the fossil record when compared with an abundance of male specimens. One theory is that antler-free female skulls were mistaken for those of horses by the people who found them, and so were discarded.

Despite its common name, Giant Irish elk, *Megaloceros giganteus* was neither exclusive to Ireland, nor related to other living elk species. Around 17,000 years ago humans saw these animals across Europe and created their own interpretations of them on cave walls, such as those at Lascaux and Cougnac. These paintings depict *Megaloceros* with speckled coats and dark shoulder hair that accentuated a distinctive hump, allowing us to speculate on their appearance.

In spite of their wide distribution, the specimen in Oxford University Museum of Natural History was found in Ireland, in a peat

bog. The Giant Irish elk arrived in Ireland from Europe around 12,000 years ago in the late Pleistocene, a period known as the Woodgrange Interstadial. This was a period of warmer temperatures which allowed the vegetation that the Giant Irish elk ate to thrive. By the Early Holocene, however, just 1,500 years later, temperatures began to cool, and *Megaloceros giganteus* became extinct.

White Watson's geological tablets

Geology is an astonishingly visual science. Understanding geological concepts, such as strata, requires an ability to perceive how they are constructed and related in three-dimensional space. To be a geologist, one must also know how to represent these concepts visually for others to understand. Over the history of geology many people have developed methods for visually representing the earth beneath our feet, most notably by using maps and geological sections. Much rarer is the use of the actual rocks and minerals in maps, sections and learning tools to represent themselves.

White Watson (1760–1835) was the grandson of the principal stonemason in the construction of Chatsworth House in Derbyshire, and heir to a family marble business. No doubt interested in geology because of his financial concerns, he was also influenced by some of the earliest publications on British geology. In 1778, John Whitehurst published *An Inquiry into the Original State and Formation of the Earth*. Inspired, Watson created a tablet of black marble featuring samples of rock inlaid as strata representing a section of mountain in Derbyshire, likely near Matlock. He hoped it would visually represent the concepts Whitehurst was proposing in his text. Whitehurst was a well-connected member of

the prestigious Lunar Society, and so when Watson created his tablet in 1785, he sent it to him. Watson was soon flooded with orders.

White Watson enjoyed great success with his original tablet design, and went on to produce a number of other types of tablets. The best known of these are the Derbyshire tablets, made from 1810, which included intricate sections of the Peak District using more detailed inlays of rocks from the region. The museum holds several of these tablets, but one of the most impressive is the section from Buxton to Chesterfield. Measuring over a metre long, it features contrasting sections of limestone with grits and shales, extending just beyond the coalfields. The brass rings on the top of these large and heavy objects show that they were meant to be hung and admired, but they are, of course, not only impressive ornaments but also examples of the fast-developing science of geology in the early years of the nineteenth century.

Stonesfield mammal jaws

From first appearances, Stonesfield is a typical West Oxfordshire village with Cotswold stone cottages and narrow winding lanes. It is unique, however, in being one of the richest sources of fossil discoveries in England. Many scientifically and historically important dinosaur remains have been uncovered in the area, and there have also been a number of other palaeontological finds, including the fossilized remains of very early mammals.

Two early discoveries of mammal fossils were both jawbones, found sometime between 1812 and 1814. They were from the Stonesfield Slate or Taynton Limestone Formation, which is from the Middle Jurassic, approximately 167 million years old. This means that the mammals lived at the same time as many dinosaurs, and made the fossils by far the oldest remains of mammals ever found at that time.

The specimens were given to Oxford academic William Buckland, who showed them to the world's leading comparative anatomist, Georges Cuvier, in 1818. Cuvier identified them as belonging to marsupials, and we now know that they belonged to an order of early mammals on the line leading to the modern-day group of mammals called Theria. This includes marsupials such as koalas and kangaroos as well as placental mammals like mice and humans.

Buckland first reported the discovery in 1824 at the same time as announcing the discovery of the famous *Megalosaurus*, the first described dinosaur, but wrote, 'the other animals that are found at Stonesfield are not less extraordinary than the Megalosaurus itself'. Buckland's announcement created a great

sensation among the scientific community of the time and it took at least twenty years to be generally accepted, with many suggesting the specimens were not as old as the Jurassic, or were from fish or reptiles. These fossils were key to the development of our understanding of the early evolution of mammals.

The great lizard

This fossilized jawbone is one of the most important dinosaur specimens in the world. It belongs to *Megalosaurus bucklandii*, a dinosaur that lived in the Middle Jurassic, approximately 167 million years ago. It would have been between eight to nine metres long and have weighed about 1,400 kilograms. It belongs to the first scientifically described dinosaur in the world.

It was found in Stonesfield, Oxfordshire, in the late eighteenth century, in a type of rock known as the Stonesfield Slate, or Taynton Limestone Formation as it is called today. This particular kind of rock was once quarried as a roofing material, though it is not actually a type of slate. In addition to building supplies, the mines at Stonesfield have also yielded more than 110 theropod bones, a subgroup of dinosaurs, from at least seven individuals. It is therefore one of the most productive British theropod dinosaur localities, and among the best localities for Middle Jurassic theropod dinosaurs in the world.

During the Middle Jurassic, Stonesfield was a shallow marine environment which saw a large input of terrestrial material, such as the fossilized dinosaurs and mammal remains found within it, likely brought out to sea via rivers. Marine animals such as fish, bivalves, gastropods, plesiosaurs and ichthyosaurs have been found there, as well as other terrestrial fossils, such as plants, insects, mammals and reptiles, including other dinosaurs.

UNDER JAW AND TEETH OF MEGALOSAURUS.

Scale ½ Inch to One Inch.

Drawn by M. Morland. On Stone by Henry Perry. Printed by C. Hullmandel.

The *Megalosaurus* jaw was first recorded in the collections of the Oxford Anatomy School at Christ Church, in an entry dated 24 October 1797, donated by Dr Christopher Pegge, Reader in Anatomy. Over a twenty-year period further specimens were found in Stonesfield, including a femur, sacrum and ribs, which were identified as coming from a giant reptile. The finds were announced together and published for the first time by William Buckland, Reader in Mineralogy, at the meeting of the Geological Society of London in 1824, the first meeting at which he was president. In this publication he named it *Megalosaurus*, making it the first ever scientifically described dinosaur. It was formally named *Megalosaurus bucklandii*, or 'Buckland's great lizard' by Gideon Mantell in 1827.

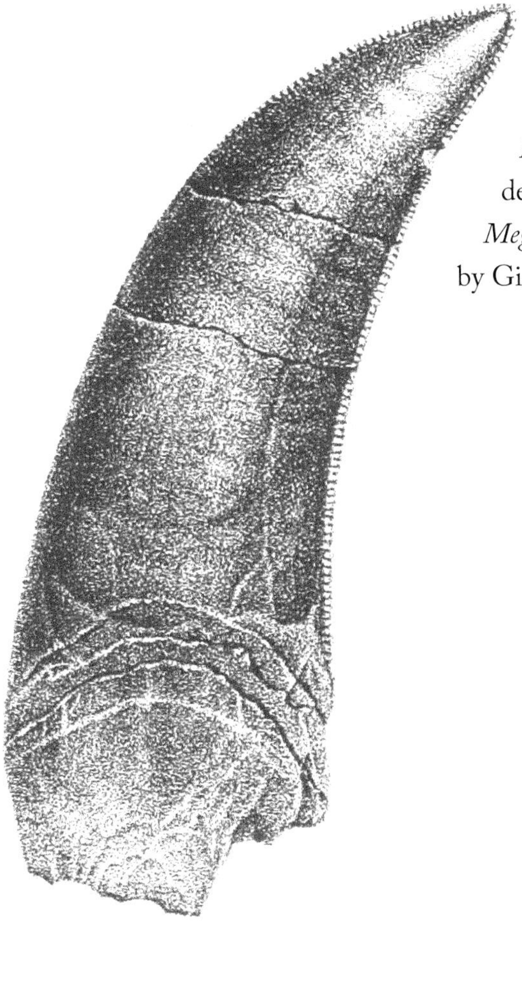

Pioneer of palaeontology

William Buckland (1784–1856) was one of the most eccentric and charismatic lecturers in the history of the University of Oxford. Appointed Reader in Mineralogy in 1813 after the resignation of John Kidd, he was the first lecturer to introduce the subject of palaeontology to Oxford students. Some of his students went on to become famous geologists and palaeontologists themselves, as is demonstrated by looking at the list of subscribers in his lecture records. These records also reveal that Buckland was able to attract the attention of many of the senior members of the university as well.

Always well attended, Buckland's lectures were animated and stimulating, and were supported by abundant use of specimen fossils and large colourful maps and diagrams to illustrate his theories. The scene of his lectures was captured in a drawing made by Nathaniel Whitlock in 1823, and demonstrates Buckland's use of objects and visual aids. One of the types of image that is featured in Whitlock's illustration, and which remain in considerable number in Buckland's archive, are geological sections. These long, colourful drawings illustrate the layers, or strata, of rock types found under the ground. Buckland often used them to demonstrate concepts such as landslips, as is seen in the illustration overleaf.

At the time Buckland was lecturing, he was based in the old Ashmolean Museum, now Oxford University's Museum of the History of Science. The Ashmolean was the only University of Oxford museum at the time, and Buckland was unofficially responsible for its natural history collections. He also contributed to the collections with rocks, minerals and

fossils he had collected in the field and that were given to him by others. All the specimens and papers were moved to this museum in 1860 after it was built to house the university's growing scientific collections and departments. It forms one of the most important founding collections of palaeontology.

Inside Mr Burchell's waggon

William John Burchell (1781–1863) was one of the greatest botanical explorers of his day. Born to a wealthy London family who owned a thriving nursery business, it is not surprising that botany became one of his lifelong pursuits, along with zoology, ethnology, art and even music. After working at Kew Gardens, Burchell began his life of travel, first to the Atlantic island of St Helena in 1805, becoming a schoolmaster and the island's botanist. The first tragic event in his life would happen just two years later when his fiancée arrived on the island; she had fallen in love with the ship's captain en route.

Five years later, he set off for Cape Town to embark on a series of explorations through Cape Province to the desolate Karoo plains. He travelled 4,500 miles, drawing accurate maps and collecting as many different plants, animals, rocks and minerals as he could. In total, he described and collected over 63,000 specimens, most of which he gave to the British Museum. He would later state that he was unhappy with the way that they cared for his collections.

While in Africa, Burchell painted one of his great artistic masterpieces, *Inside Mr Burchell's Waggon*. Some of the specimens to be seen in Burchell's painting are in Oxford University Museum of Natural History's collections. They include a tortoise that he collected on 10 October 1814, the tusk of a

hippopotamus and the molar tooth of an elephant. There are also a number of insects, depicted in the top right corner. The painting was highly regarded for its artistic quality and was hung in the Royal Academy as part of its 1820 exhibition. It is now also housed in the museum's collections.

Although the University of Oxford awarded him an honorary degree in 1834, Burchell's work was not widely recognized. Disillusioned and unhappy, he committed suicide in 1863. In 1865, his sister donated his remaining plants to Kew Gardens, and all his other specimens to Oxford University Museum of Natural History. They include many type specimens of species that were new to science.

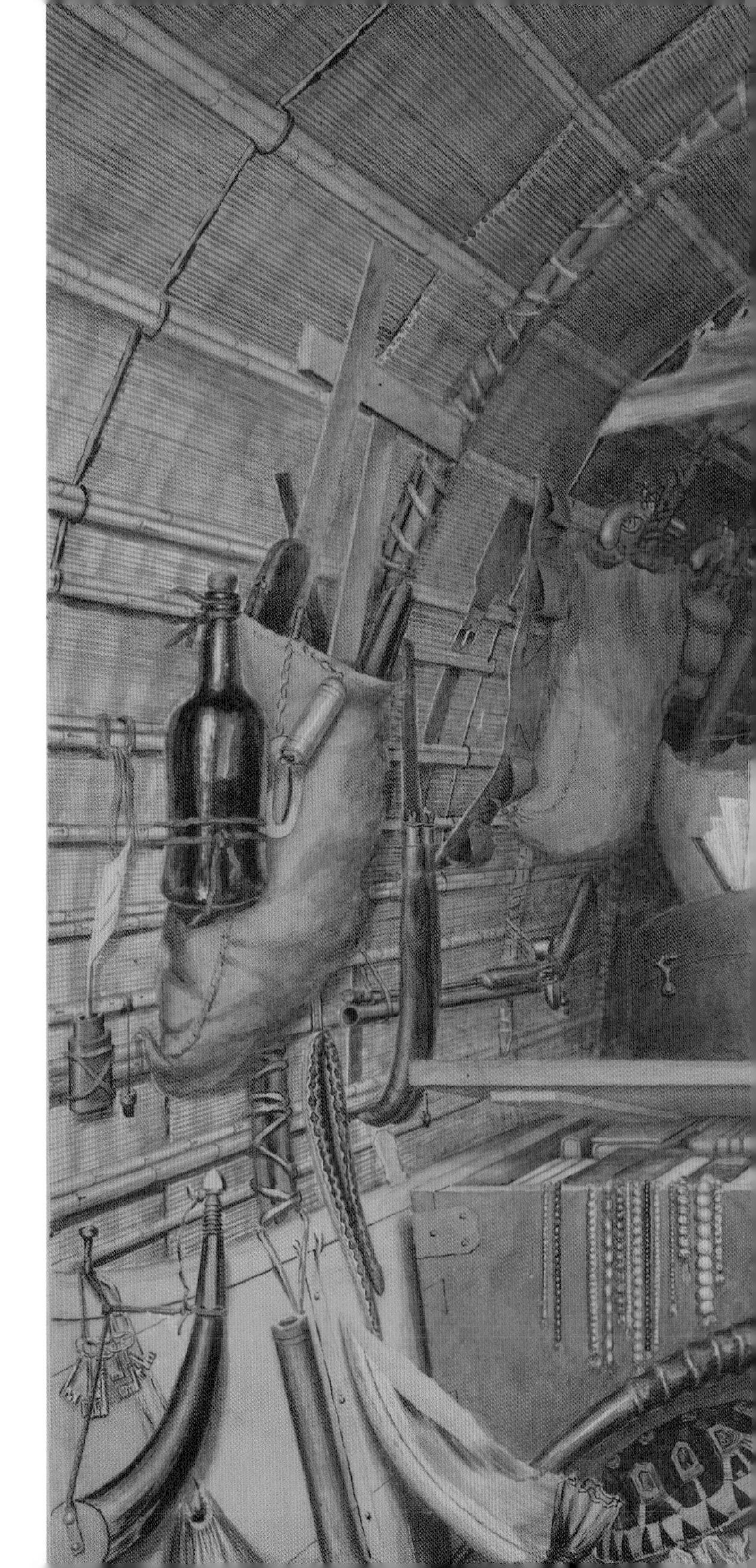

Smith's geological map

The Industrial Revolution marked a time of great change in Britain, both for the landscape and for the lives of its inhabitants. People were moving to the cities with hope of employment resulting from technological advances, which resulted in new transportation links being created: first came canals which were then quickly followed by railways. The demand for coal was unprecedented, and the ability to locate it and efficiently mine it was a priority.

This map, entitled *A Delineation of Strata of England and Wales*, was published at the climax of this rapid expansion. Its creator, William Smith (1769–1839), was a working man, employed as a land surveyor and agricultural engineer. His passion and astute observation of geology led him to make an important discovery about the predictability of the layers of rock found beneath the earth. He was the first to determine and record that rock layers could be identified by the fossils found within them and that these layers formed a pattern that extended across the country.

Before the publication of Smith's great map in 1815, knowledge of geology over large distances was not wholly understood. Mining was already an established industry, but often mineral deposits were found by accident. With the increased demand for coal and other minerals, the ability to systematically predict where and how far down these types of materials could be found was crucial. Smith conceived the idea of the map in 1801, but it took fourteen years before a completed copy would be printed. He gathered the data needed to create the map entirely on his own, while travelling the country to fulfil engineering contracts. By the time the map was printed in August 1815, many of its subscribers had lost interest, no

longer had the funds available to buy it or had even died. Also, the Geological Society in London, which had earlier rejected Smith's application for membership, then produced its own map, largely plagiarizing his work. Smith had invested his life's savings into the work and found himself in debtor's prison within a few years. He would only receive recognition for his incredible accomplishment in the last few years of his life.

However, the fundamental methods that Smith devised in order to create his geological map are still used by geologists today to map the entire planet, and even other planets in our solar system.

Hyenas in Yorkshire

In 1821, a cave littered with unusual broken animal bones was discovered by labourers working in a roadside quarry at Kirkdale in Yorkshire. News of the unexpected find of what appeared to be remains of exotic animals such as elephant, rhinoceros, hippopotamus and hyena, among others, was sent to William Buckland (see p. 63), Reader in Geology at Oxford, who promptly visited in November 1821.

At this time, it was generally assumed that finds such as these must have been the result of the biblical flood, with the remains of drowned animals being carried great distances from their original habitat and deposited when the waters receded. Buckland, however, using developing scientific knowledge and method, came to a different conclusion. He demonstrated that the significant quantity of hyena remains and the splintered states of all the bones could only mean one thing: that the cave had actually been inhabited by hyenas before the flood. The hyenas, he suggested, had ranged over the surrounding countryside, killing prey and dragging it back to their lair. Since several of the species present only live in tropical regions he concluded

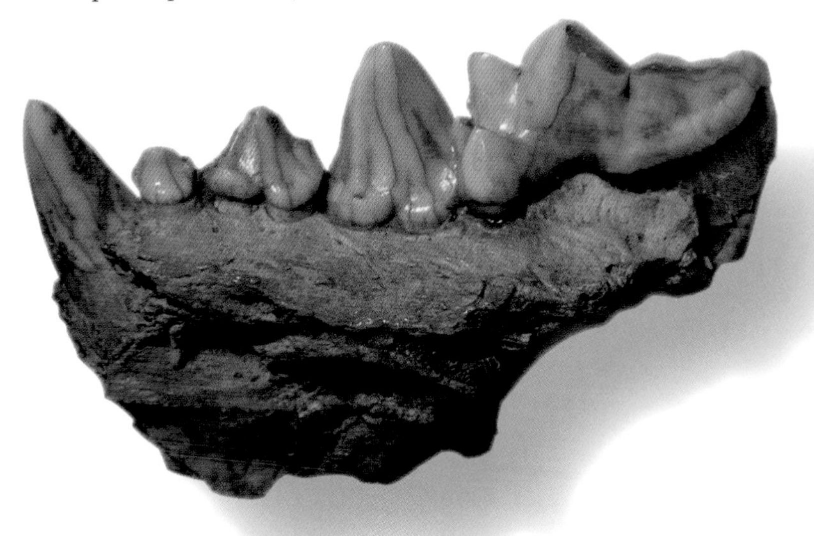

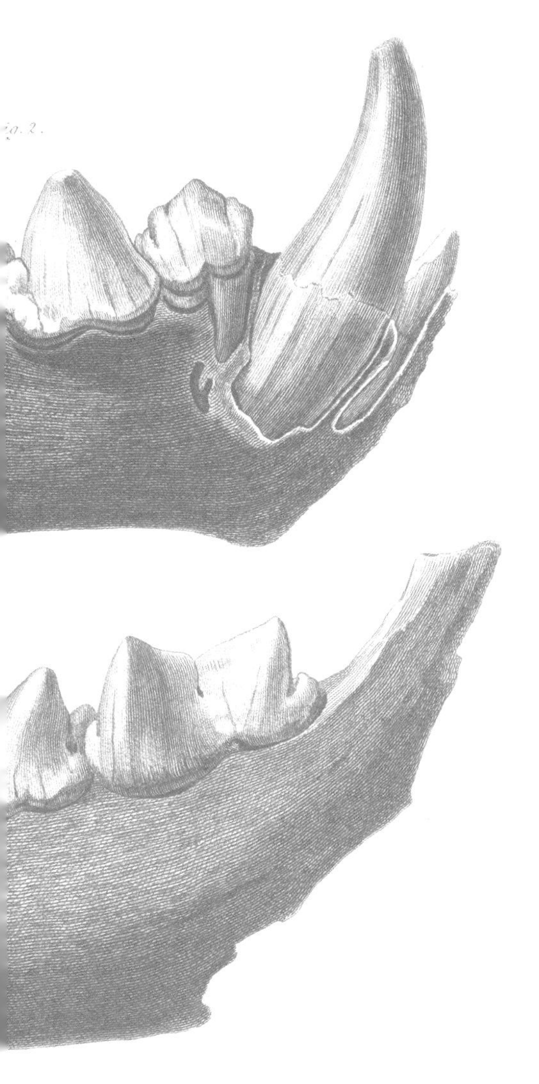

that Britain must once have been much warmer than it is today.

To test this theory, Buckland conducted a number of experiments. He gave an ox shinbone, acquired in the Oxford market, to a hyena passing through Oxford with Mr Wombwell's travelling menagerie. The living hyena produced gnaw marks on the bone that were identical to those on a comparable bison shinbone from Kirkdale.

Buckland also noticed some small balls of white material lying among the bones and teeth, and wondered if they might be fossilized faeces from the hyenas. He referred to them as *Album Graecum*, an old apothecary's term for a type of dog faeces which turns white on exposure to air. This 'drug', used to treat colic, dysentery, scrofula, ulcers and quinsy, was obtained by feeding half-starved dogs with bone fragments. When the protein in the bone had been digested and absorbed, it left a phosphate-rich faecal pellet.

To test his hypothesis, Buckland sent some of the Kirkdale material to chemist William Hyde Wollaston (1766–1828). Wollaston showed the specimens to the Menagerie Keeper at the Exeter Exchange, who immediately noted their similarity to the droppings of the spotted hyena (*Crocuta crocuta*). Wollaston's analysis found the coprolite 'to be composed of the ingredients that might be expected in faecal matter derived from Bones' (Buckland, 'Account… in the Year 1821').

We now understand that the Kirkdale cave was occupied by hyenas approximately 120,000 years ago, during a warm spell between glaciations.

The Red Lady of Paviland

The 'Red Lady' was found in Paviland Cave on the Gower Peninsula, Wales, in 1823 by Oxford Reader in Geology William Buckland. The find comprised the partial skeleton of a single individual stained a red hue using ochre, a naturally occurring pigment containing iron oxide, as well as ochre-stained mammoth ivory and bone ornaments, including a pendant. It was the discovery of these particular artefacts that led Buckland to believe that the individual was female, giving her the name the 'Red Lady'. It was the position of the remains and the associated items, found laid out in a shallow grave, that led to the conclusion that this was a burial. Buckland understood that this find was something remarkable.

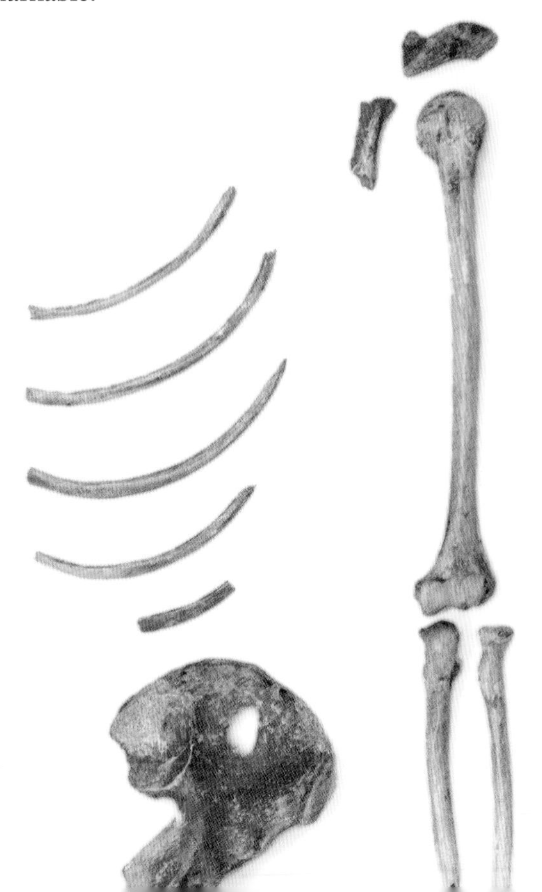

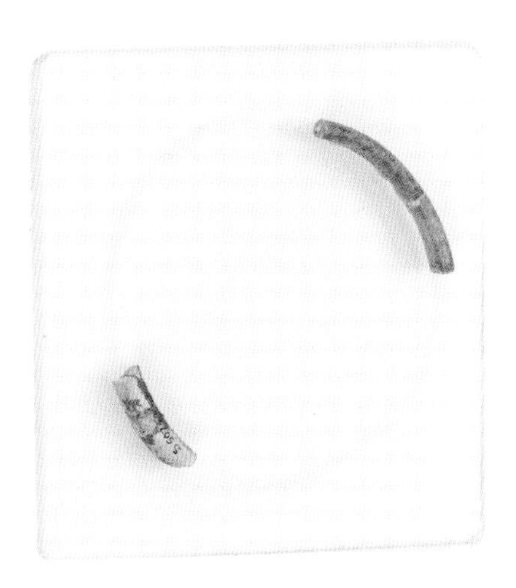

Since the discovery of the 'Red Lady' much has been uncovered of its circumstances. Most importantly, we now know that 'she' is in fact a man. The skeleton is of a young man, with recent radiocarbon dating indicating the remains are approximately 33,000 years old. This makes the burial the oldest known of *Homo sapiens* (modern human) in Western Europe; it is now an iconic find of the British Palaeolithic period. We know the ochre found on the skeleton and artefacts was likely sourced from veins of iron ore exposed at Mumbles Head, 18 miles to the east of Paviland Cave. This means that the pigment was deliberately applied to the bones of the deceased, implying that the body was stored and treated after death. There is another theory that the pigment was applied to clothes before burial and was transferred as they decomposed.

The exact circumstances of the development of burial customs in human culture are still unknown, but the 'Red Lady' raises important questions for understanding this fundamental characteristic of human society.

The Oxfordshire dinosaur

In 1825 at Chapel House, near Chipping Norton in Oxfordshire, the remains of a giant dinosaur were found by John Kingdon, a local man. These included the fossilized bones of the neck and foot, and belonged to a creature that would later be named *Cetiosaurus*, or 'whale lizard'. It was the first of its kind ever found.

Cetiosaurus was one of the first sauropod dinosaurs scientifically described, sauropods being the largest land animals to have ever lived on the planet. This sauropod lived in the middle part of the Jurassic period, around 166–168 million years ago and would have weighed an incredible 27 tonnes, about four to six times more than an African elephant. Sauropods were also herbivores, eating plant material only. They had long necks and small heads in proportion to their bodies, and they walked on four legs.

Richard Owen (1804–1892), a famous nineteenth-century palaeontologist, named *Cetiosaurus* in 1841. Since the first find in 1825, several other examples of the genus had been found in Oxfordshire,

Northamptonshire, Gloucestershire and even Yorkshire. Owen believed that the huge creature must have been a giant marine reptile that ate crocodiles and plesiosaurs. It was not until thirty years later, and based on yet more discoveries, that the Keeper of the University Museum, John Phillips, named *Cetiosaurus* as a new species, *Cetiosaurus oxoniensis*. He also suggested that it was an amphibious reptile rather than marine, and formed the theory, based on its dentition, that it was probably a plant eater.

Our understanding of *Cetiosaurus* and sauropods in general is much greater than it was at the time of Owen and Phillips, but they still remain the largest dinosaurs we know of that lived in the United Kingdom.

The Corsi Collection

The Corsi Collection is an extraordinary and beautiful assemblage of 1,000 polished slabs of decorative stones, all similar in size. It was created by Italian lawyer Faustino Corsi (1771–1846) from stones collected across the city of Rome and from further afield, in locations such as Russia, Afghanistan, Madagascar and Canada. Corsi also received gifts of stone from early patrons of art and archaeology, including the Duke of Devonshire, William Cavendish, who contributed a fine collection of rocks and minerals from Derbyshire, his home county.

As well as collecting, Corsi also published a catalogue of his collection in 1825, *Catalogo ragionato d'una collezione di pietre antiche*. The work provided names and descriptions of his specimens and attempted to organize them using geological principles. It also offered advice on how one could identify similar specimens, and locations to both find and view examples of the particular type of rock or mineral he described. This and Corsi's other book, *Delle pietre antiche*, were among the most important reference works on Roman decorative stones for nearly 140 years.

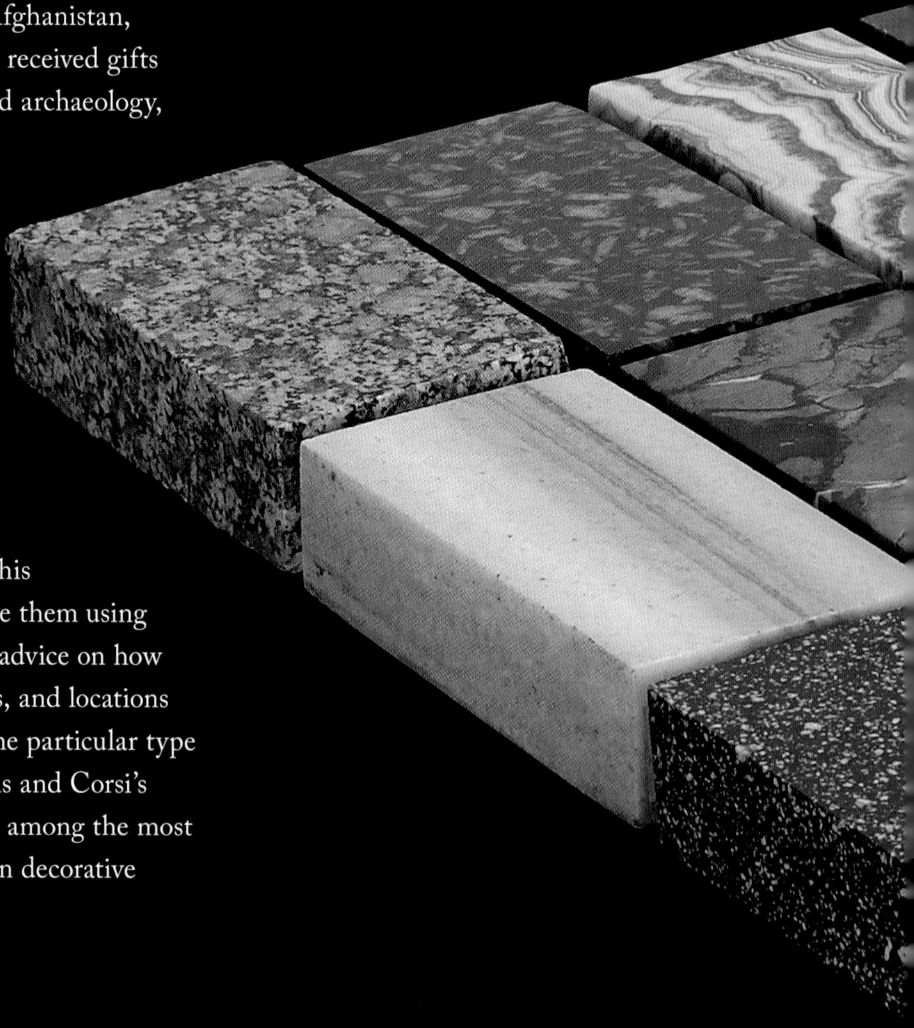

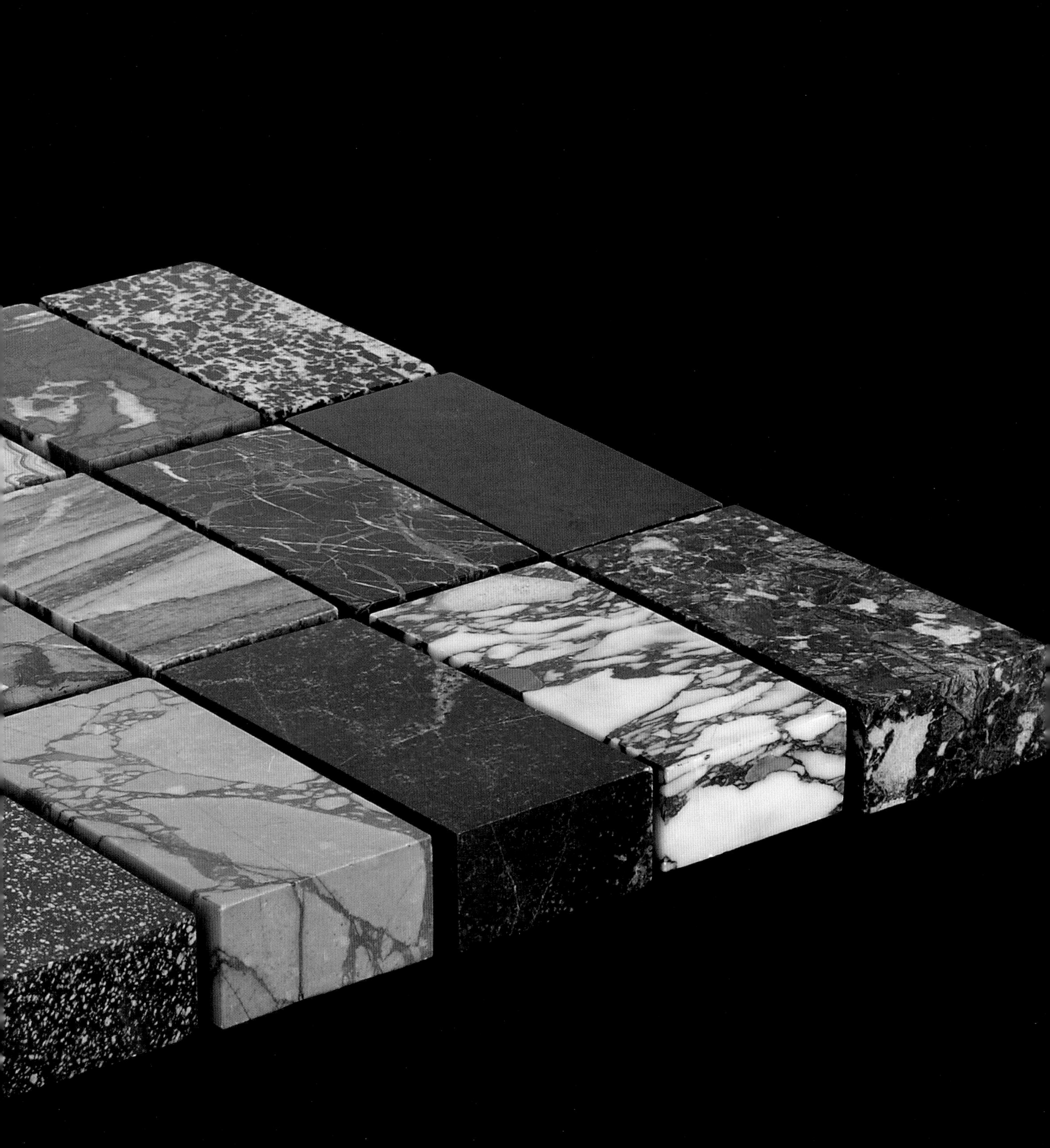

The collection was purchased from Corsi and presented to the University of Oxford by Magdalen College student Stephen Jarrett in 1827. It was later moved to the museum after its construction in 1860 as a repository for the university's scientific collections. It remains an exceptional reference collection for identifying and researching stones used to ornament buildings, furniture and artefacts from the Byzantine to Baroque periods, and through to modern times. It is used by a wide variety of researchers, conservators, artists and those in the antiques trade, as well as archaeologists and geologists. It is also available to search and browse online in its entirety.

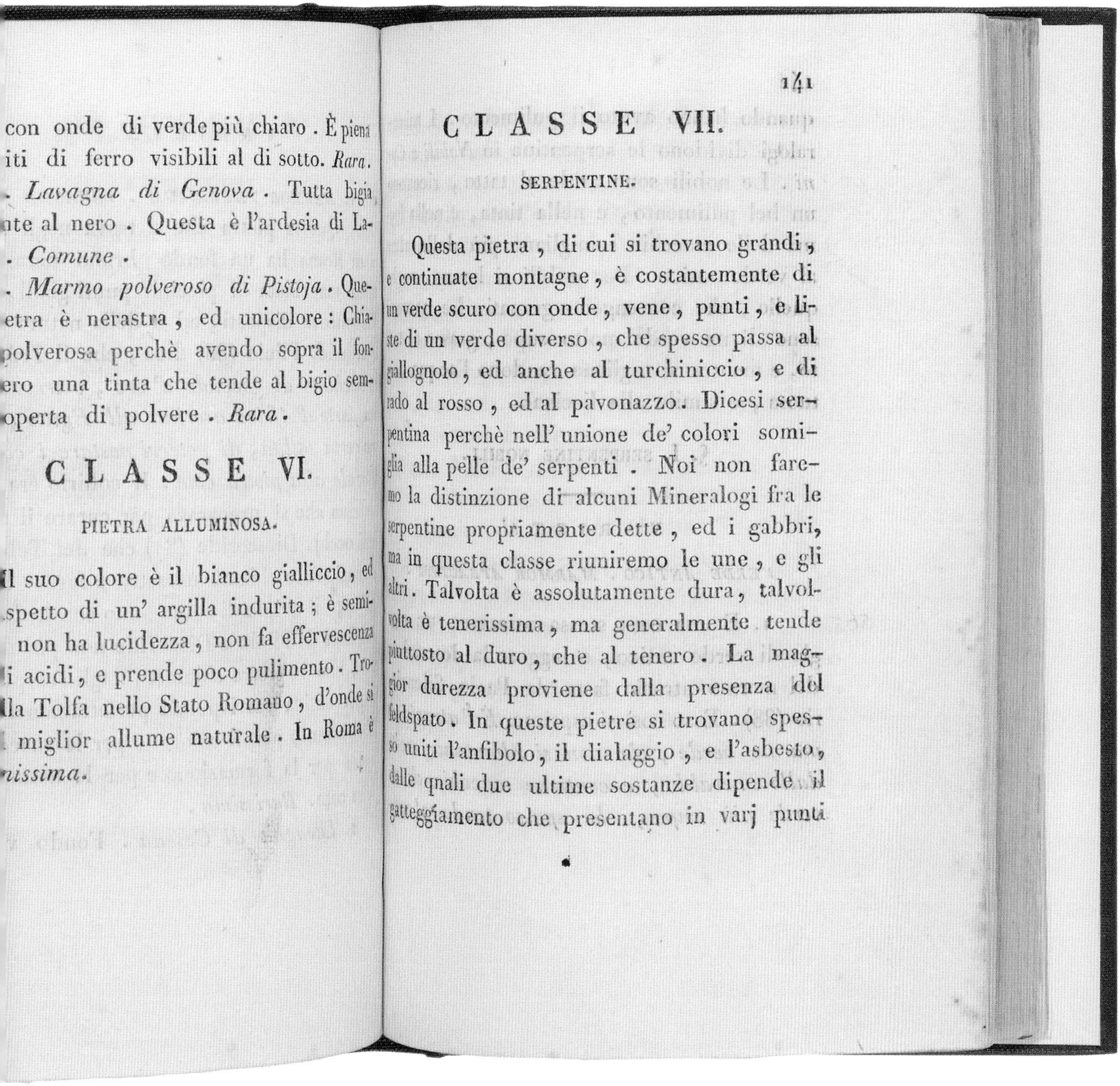

con onde di verde più chiaro. È piena
iti di ferro visibili al di sotto. *Rara*.
. *Lavagna di Genova*. Tutta bigia
nte al nero. Questa è l'ardesia di La-
. *Comune*.
. *Marmo polveroso di Pistoja*. Que-
etra è nerastra, ed unicolore: Chia-
polverosa perchè avendo sopra il fon-
ero una tinta che tende al bigio sem-
operta di polvere. *Rara*.

CLASSE VI.

PIETRA ALLUMINOSA.

Il suo colore è il bianco gialliccio, ed
spetto di un' argilla indurita; è semi-
non ha lucidezza, non fa effervescenza
i acidi, e prende poco pulimento. Tro-
lla Tolfa nello Stato Romano, d'onde si
miglior allume naturale. In Roma è
nissima.

CLASSE VII.

SERPENTINE.

Questa pietra, di cui si trovano grandi, e continuate montagne, è costantemente di un verde scuro con onde, vene, punti, e liste di un verde diverso, che spesso passa al giallognolo, ed anche al turchiniccio, e di rado al rosso, ed al pavonazzo. Dicesi serpentina perchè nell' unione de' colori somiglia alla pelle de' serpenti. Noi non faremo la distinzione di alcuni Mineralogi fra le serpentine propriamente dette, ed i gabbri, ma in questa classe riuniremo le une, e gli altri. Talvolta è assolutamente dura, talvolvolta è tenerissima, ma generalmente tende piuttosto al duro, che al tenero. La maggior durezza proviene dalla presenza del feldspato. In queste pietre si trovano spesso uniti l'anfibolo, il dialaggio, e l'asbesto, dalle qnali due ultime sostanze dipende il gatteggiamento che presentano in varj punti

The lost volcanic island

In the middle of the Mediterranean Sea, about thirty kilometres west of Sicily, there is an underwater volcano that occasionally erupts, creating an island. One such eruption occurred in July 1831. It was seen on 1 August by H.F. Senhouse, captain of the Royal Navy ship *St Vincent*. He named this previously unknown landmass Graham Island after Sir James Graham, the First Lord of the Admiralty, and claimed it for the United Kingdom. Rival claims were also made for this new island, with the Bourbons naming it Ferdinandea after Ferdinand II of the Two Sicilies, and France calling it Giulia because it appeared in the month of July.

Graham Island reached a height of approximately sixty metres and had a circumference of about four kilometres. It was situated on an important shipping route and became a tourist attraction as well as an object of territorial dispute. The island was formed from tephra, loose fragments of basaltic lava and ash. Such unconsolidated material made it particularly vulnerable to erosion by the sea and, by December 1831, the island had disappeared completely below the waves, leaving the issue of sovereignty unresolved.

A number of drawings and paintings of the island are known, but samples of the volcanic rock itself are rare and irreplaceable. Four of these are in the extensive geological collections made by Professor Charles G.B. Daubeny (1795–1867) held in Magdalen College until they were transferred to the museum in the 1950s. Daubeny described the island in the 1848 second edition of his book, *A Description of Active and Extinct Volcanoes*, and cites the reports to the Royal Society of Dr John Davy, who visited the island in August 1831 and is likely to have collected the specimens in the museum.

Three of the specimens are samples of volcanic lava, the third is a box of volcanic sand. They enable scientists to learn about volcanic activity that normally only occurs below the sea.

Bless the baby! what a Walley he have a-made. !!

Cause and Effect

Cause and effect

This satirical illustration, drawn by Henry De la Beche around 1830, is, it first appears, of a young boy urinating to make a stream. In addition to its intended humour, it is in fact a commentary on a contemporary geological theory, with the caption reading 'Bless the baby! What a Walley he have a-made!!!' It is part of the William Buckland Collection and it is believed that the boy represented was Buckland's son Frank.

De la Beche, in addition to being an eminent geologist, was infamous for his satirical cartoons commenting on the science of the day. His diaries are full of sketches about contrasting and developing geological theories, and Charles Lyell, another important contemporary geologist, featured regularly.

This illustration was done at about the same time that Charles Lyell published his seminal work *The Principles of Geology*, in 1830 (see p. 198). In it, Lyell proposes that most valleys were formed over very long periods of time through erosion caused by running water, such as rivers. This theory was in contrast to the widely held theory at the time that valleys were the result of geological events such as earthquakes, or were caused by floods. This supposition was part of Lyell's theory of uniformitarianism, which sees the formation of the Earth's crust as a result of slow, continuous and uniform processes. On a larger scale, beyond just the field of geology, it is the understanding that the natural laws of the universe are constant and apply everywhere.

While Lyell's theories have merit, our understanding of how these processes work is more complicated than he originally set out. For example, we now know that the particular type of valley illustrated here was the result of glacial movements – a theory that had not then been developed.

Captured in squid ink

Mary Anning is one of the most famous early female fossil hunters, but her contemporary, Elizabeth Philpot, was equally influential and active, though less known. Just like Anning, Philpot was well connected with many of the early palaeontologists across the country, many of whom were eagerly seeking the abundant fossil remains being uncovered along the Dorset coastline now known as the World Heritage Site the Jurassic Coast.

Two such palaeontologists were William Buckland and Louis Agassiz. Agassiz visited Elizabeth and her sisters in 1834 to see their impressive collection. He was struck by their extensive knowledge, citing Elizabeth in his seminal work *Recherches sur les poissons fossiles* and naming a species, *Eugnathus philpotae*, after her. It was rare in that period for women to be properly acknowledged for the scientific work they did and the contributions they made, and the frequent correspondence between Philpot and the contemporary scientists of the day was a real indication of her influence and the respect that was held for her work.

It is in her correspondence with William Buckland, found in the archives held in Oxford University Museum of Natural History, that another interesting discovery she made is represented. In 1826, Mary Anning discovered an unusual material in the chamber of a fossilized belemnite, an extinct squid-like cephalopod, that appeared to be ink. She showed it to Philpot, who knew that by adding water she could revive the ink, making it useable. Philpot used it to illustrate the remains of an ichthyosaur she had recently found in a letter to Mary Buckland, dated 9 December 1833; she also made a copy of the painting. It was a technique that caught on, and artists in the region began using fossilized ink in their work, a practice that is still used to this day.

A Jaw of the Ichthyosaurus communis
from the lias, Lyme Regis.
Drawn with colour prepared from
the fossil Sepia cotemporary with
the Ichthyosaurus.

Half the
natural size

The Beringer Lying Stones

Fakes and forgeries are not unknown to those who work in museums and galleries, but most often they are copies of the work of famous artists, such as paintings or sculptures. Natural history museums, however, are not strangers to frauds, and examples of one of the most famous cases of palaeontological forgery are housed in Oxford University Museum of Natural History, now as relics of the hoax rather than as an example of the fossils they claimed to represent.

Dr Johann Beringer (1667–1740) taught natural history at the University of Würzburg in the early eighteenth century. He was known to have an impressive collection of fossils and in 1725 obtained what he believed to be exceptional examples of fossilized life forms unlike anything he had seen before. In particular, they appeared unique because they were all relatively the same size, with creatures represented in their entirety and in natural poses, rather than on their side or back. They were also very smooth on the front sides, as though they had been polished, and appeared to be in relief, not showing any signs of compaction normally seen on fossils. Despite these unusual and suspicious features, Beringer remained convinced of their authenticity and began a publication outlining the collection.

The fossils were acquired from three youths Beringer employed as

fossil collectors, Christian Zänger and brothers Niklaus and Valentin Hehn. Between June and November 1725 they 'collected' approximately 2,000 of the stones. The fossils were in fact manufactured and polished by J. Ignatz Roderich, Professor of Geography, Algebra and Analysis at Würzburg University, and Georg von Eckhart, Privy Councillor and Librarian to both court and university. At a later interrogation, after Beringer brought charges against the two men, a third party admitted to hearing the men discuss how they wished to discredit and humiliate Beringer for being arrogant.

The museum holds two of the mysterious Beringer Lying Stones, as they are known from a direct translation of the German. They were acquired by William Buckland from Professor Ludwig Rumpf in 1835. Both are inscribed by Mary Buckland, William's wife, who wrote, 'Artificial fossils See Beringer who died of a broken heart from mortification at…' The rest is illegible.

An ichthyosaur's last meal

The ichthyosaur, resembling a cross between a fish and a dolphin, is an extinct marine reptile and iconic sea creature that lived at the time of dinosaurs in the Mesozoic period, approximately 250 to 90 million years ago. The museum holds the fossilized remains of a number of species of ichthyosaur (another is shown overleaf), but one in particular has an interesting connection to a very famous fossil hunter.

This specimen was found by Mary Anning (1799–1847) at Lyme Regis. It contained the remains of what appeared to be fossilized food in its abdominal cavity, including large fish scales and spines identified as belonging to the Lias fish *Pholidophorus*. It was given to William Buckland, Reader in Geology at Oxford, sometime after 1836 by the Rt. Hon. William Willoughby Cole, who had bought the specimen from Anning. The specimen was then given to Buckland because of his academic interest in the digestive stuff of Jurassic creatures, in particular his fascination with fossilized poo or 'coprolites', from the Greek for 'dung stones'.

In 1829 Buckland identified potato-shaped stones found at Lyme Regis as the fossilized droppings of ichthyosaurs, based upon their co-occurrence with these marine reptiles and also their contents of bones and scales of fish, and bones of smaller ichthyosaurs. Buckland had also famously studied the coprolitic matter of extinct hyenas found in Kirkdale Cave, Yorkshire.

The theme was taken up pictorially by Henry Thomas De la Beche (1796–1855), whose watercolour sketch *Duria Antiquior: A More Ancient Dorsetshire* (1830) shows a scene from the Lower Lias sea in which coprolites float down onto the seabed as the marine reptiles and fish go about their daily business of eating and being eaten. De la Beche then paid the artist Georg Scharf to create lithographic prints of the sketch, which were sold in aid of the Anning family, who made little money from their remarkable fossil finds. There is a copy of this amusing illustration in the museum's archives.

Though this specimen was originally identified as *Ichthyosaurus communis*, it has recently been re-identified as a juvenile of *Ichthyosaurus anningae* by the species authors Dean Lomax and Judy Massare in 2015. The species is named after Mary Anning, and the specimen shown here is one of the few examples found by Mary herself.

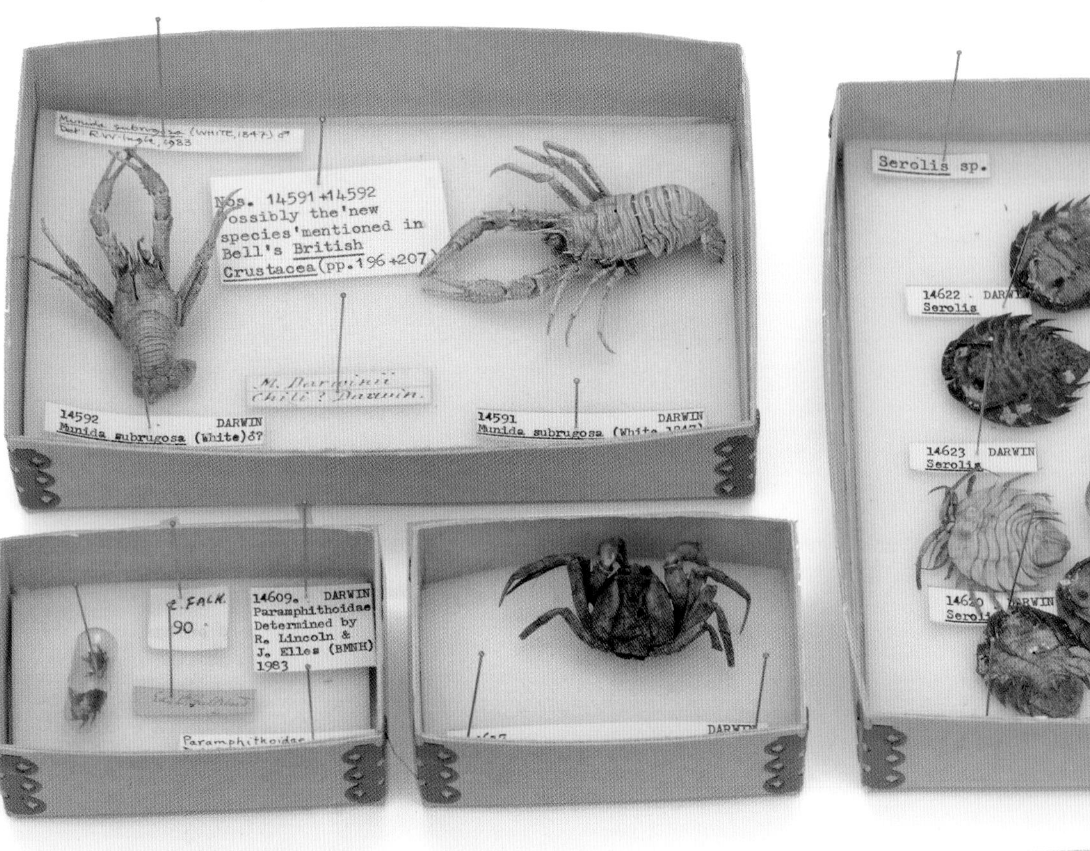
Nos. 14591 +14592
Possibly the'new species'mentioned in Bell's British Crustacea(pp.196 +207)
14592 DARWIN
Munida subrugosa (White)♂?
14591 DARWIN
Serolis sp.
14622 DARWIN
Serolis
Serolis
14623 DARWIN
Serolis
14626
Serolis
DARWIN
14609. DARWIN
Paramphithoidae
Determined by
R. Lincoln &
J. Elles (BMNH)
1983

GRIMOTHEA
14562 DARWIN
Munida gregaria (Fabricius,1793) ♀?
14563 DARWIN
Munida gregaria (Fabricius,1793) ♀

14539
14539 DARWIN
Cancer polyodon Poeppig,1836 ♂

Darwin's crabs

Charles Darwin (1809–1882) read for his undergraduate degree in Cambridge, graduating in 1831. During his time there he regularly discussed the natural world with lecturer and botanist John Henslow, forming a lifelong friendship. It was Henslow who recommended Darwin for the post of naturalist and captain's companion on board HMS *Beagle* after turning down the post himself. Darwin and Henslow would continue to correspond throughout the five-year voyage and Henslow's work was one of the many influencing factors on Darwin's own later publications.

Darwin not only made ecological and geological observations during the voyage of the *Beagle*, but also amassed a vast collection of many kinds of natural history specimens. On his return to England these were entrusted to various scientists for study. The Crustacea, or crabs, shrimps and lobsters, were sent to Thomas Bell, who is now better known for his work on turtles and tortoises. Bell was widely thought of as the expert on the group at the time but he was slow to work on the material and the specimens were never properly described. They remained in Bell's cabinets for many years until, in 1862, they were purchased by the first Hope Professor of Zoology, John Obadiah Westwood, on behalf of the University Museum, along with other specimens and materials.

Of the Crustacea that remain in the museum collection, some are stored dry, such as the specimens illustrated here, whilst others are preserved in spirit, although it is clear from Darwin's notebooks that originally everything was stored in spirit. Many of the surviving specimens have either numbered labels attached to them in handwriting that has been ascribed to Syms Covington,

Darwin's servant on the *Beagle*, or numbered metal tags. These numbers correspond with those listed in the 'Catalogue for Specimens in Spirit of Wine', which is a charmingly titled and chronological listing of specimens collected throughout the voyage. Of the 230 crustacean specimen lots listed in this catalogue all are marked in pencil with the letter *C*, presumably to allow a separate list to be copied out. Only 110 of these lots can be traced and reliably identified today as those collected by Darwin, though further specimens can be tentatively linked to him through more circumstantial evidence.

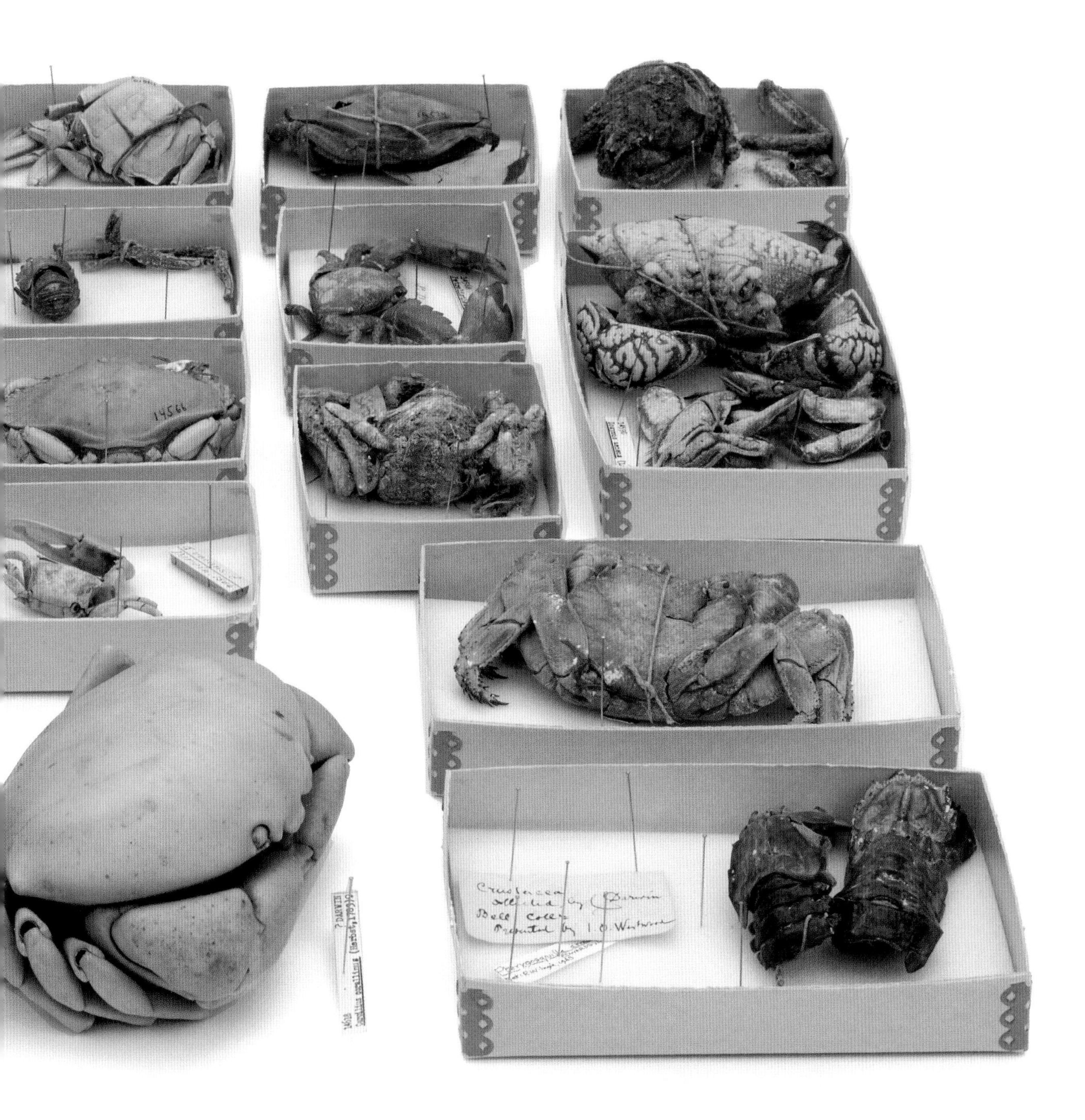
14566
Crustacea
Bell Coll.
Presented by I. O. Westwood

SCALE FOR THE THREE INCH MODEL
SCALE FOR ALL THE REST OF THE THREE INCH MODELS

Sopwith's geological models

These attractive wooden blocks belong to a series of geological models that were designed to illustrate the nature of the Earth's crust, in particular stratification and the effects created by faults and mineral veins. They were created by builder and later land surveyor, Thomas Sopwith (1803–1879), in 1841. Few Sopwith models exist today and this particular set was given to William Buckland, Reader in Geology at Oxford, by Sopwith himself, who acknowledged Buckland's help with the selection of models that could be sold together as a set.

The models came in sets of twelve and were constructed from hundreds of separate pieces of wood that were laminated and joined together, and then carved by hand. They were intended as educational tools to teach fundamental geological concepts, with each shade of wood representing a different stratum or layer of earth. They were individually constructed to represent different geological features, based on various locations in the North of England, predominantly mining areas.

The Sopwith models were sold in a box designed to look like a book, allowing customers to conveniently store their models in an attractive object. The book shape was also used by Sopwith to illustrate the concept of strata, or the layers of tilting rock; he compared the concept visually to books that had fallen over on a bookshelf. In the second set, released in 1875, the models were reduced in number to six and accompanied by a small publication entitled *Description of a Series of Geological Models.* The craftsmanship and labour-intensive method employed in making them meant that the cost of the models was considerable, though sales were good.

Vesuvius erupting

This dramatic painting, a gouache on paper, depicts Mount Vesuvius in the Gulf of Naples, Italy, violently erupting. It is a common scene featured in hundreds of paintings done in the region in the late eighteenth and early nineteenth centuries. The use of light and shadow was typical of the style, though not all such paintings were done at night. The artist of this particular painting is unknown, but the image has been compared to works by the obscure Italian artist Enrico La Pira. The work is dated 1845.

Mount Vesuvius is one of the most famous volcanoes, most commonly remembered for its eruption in 79 CE, which completely buried the Roman cities of Pompeii and Herculaneum. It is, however, still one of the most dangerous active volcanoes in the world and has erupted many times since, though with less catastrophic results. It is a stratovolcano, meaning it has been built up over time into a conical shape made up of many layers of volcanic materials such as lava and ash. It also has a caldera, or crater, in the middle, caused by the collapse of an earlier and originally much higher rock structure. Vesuvius was formed, like many volcanoes, by the collision of two tectonic plates. This allows hot magma from within Earth's mantle to make its way to the surface.

Vesuvius erupted several times in the nineteenth century, sending artists and geologists from around Europe flocking to the scene. This painting is part of a set of twelve depicting Mount Vesuvius erupting, found in the William Buckland Print Collection. How and from whom Buckland received these paintings is unknown, but they would have likely been used by Buckland to gain a further understanding of the geology of volcanoes and volcanic eruptions, known as volcanology.

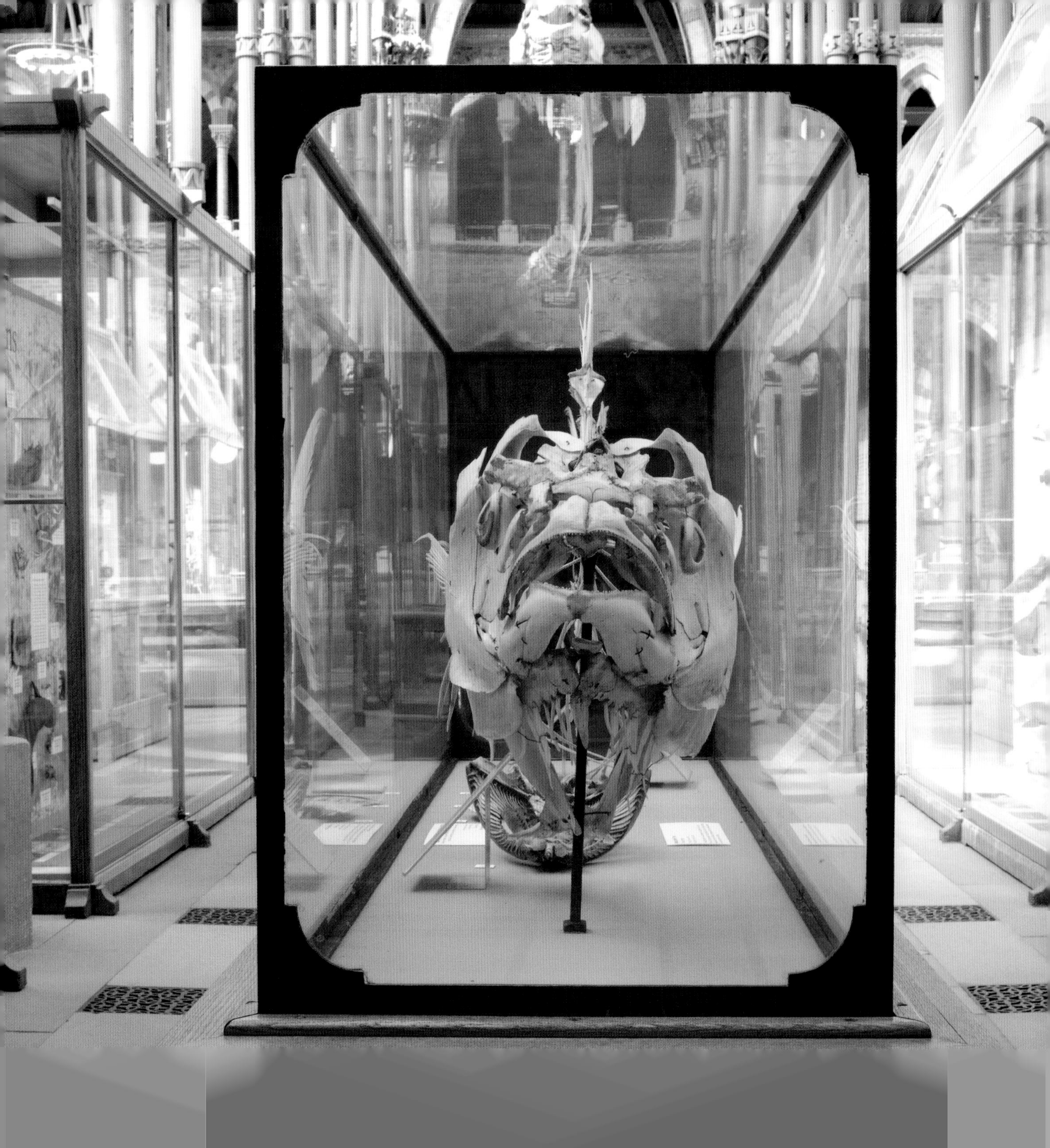

The tunny

Salted and packaged in an eight-foot-long wooden box addressed to 'Dr. Acland, Oxford', the tunny, now better known as the Atlantic bluefin tuna (*Thunnus thynnus*), began its voyage to the University Museum from Madeira, accompanied by the museum's founder and principal collector, Henry Acland, early in 1846.

The journey was an eventful one for both man and fish. An awful storm blew up in the Bay of Biscay on 9 or 10 January, and the crew and passengers alike fixated their fears on the large crate, and assuming Acland was a medical doctor imagined the contents to be the corpse of a patient. Being superstitious, they linked this in their minds to the bad weather they had encountered, no matter how strongly the crate's owner argued otherwise.

Their unrest rose to such a fervour that there was talk of a mutiny if the box was not disposed of. In order to quell the unrest the captain was forced to approach Acland and tell him that the ill-omened box was to be thrown over the side. Acland threatened legal proceedings so as to preserve the specimen but the crew and passengers were not convinced, and an unhappy truce fell into place with crew and passengers alike ignoring or refusing to speak to him for the rest of the voyage.

Faced with this situation, Acland allowed the ship's carpenter to unscrew the tunny's crate so as to reveal the contents and the truth of his words to all on board. The rest of the voyage seems to have passed in relative peace until 13 January, when, after running ten miles off course, the ship hit a reef and was grounded off the Dorset coast.

Thankfully a small boat was able to take all passengers safely to shore. The sailors, possibly motivated by feeling of guilt and remorse for the difficult time they had given Acland over the contents of his crate, worked double tides to rescue the tunny from the ship. It was later delivered in perfect condition to Oxford, where it has since led a more peaceful existence.

When the museum was opened in 1860 the articulated skeleton of the tunny was one of the first objects to be put on public display in the centre court, where it has remained to this day. It is believed to be the longest continually displayed specimen in the museum.

Cornish chalcotrichite

One of the most impressive collections of minerals in the museum was amassed by Dr Richard Simmons, FRS, a classic nineteenth-century gentleman of leisure. Trained as a physician, he had the good fortune to retire at a very young age after inheriting his father's estate (his father had been physician to the 'mad king', King George III). Simmons spent the rest of his life collecting minerals and fine works of art.

A number of Richard Simmons' minerals were collected from mines and quarries across the United Kingdom. Since the mining industry in Britain is now much smaller than it was during the height of the Industrial Revolution, most of these mines are now closed and inaccessible to researchers and collectors. This makes the specimens in his collection of particular scientific importance, as many are now rare.

One such specimen is a beautiful example of chalcotrichite from Gunnislake, Calstock, in Cornwall. Chalcotrichite is a variety of the mineral cuprite which has slender hair-like crystals, very different from the typical cubic, octahedral or dodecahedral crystal shapes shown by the crystals of this mineral. It is formed in the oxidized zones of copper ore deposits and Cornwall was one of the world's major suppliers of copper ore in the late nineteenth century. This specimen is believed to be the finest example of its type from the area.

Richard Simmons bequeathed his impressive and scientifically important collection of minerals to the University of Oxford in 1846. To this day, they represent some of the museum's most beautiful and valuable mineral specimens.

Temperance & trilobites

Dr Ralph Barnes Grindrod (1811–1883), an unorthodox doctor and vocal advocate of the temperance movement, was an unlikely man to establish a Museum of Geology and Natural History. His museum, however, was declared the 'finest private collection of its kind in all of Europe, if not the world' by *The Templar* in 1874. The location of the museum was also unconventional. Established at Townsend House in Great Malvern, which also served as a clinic practising 'water cures', or hydrotherapy (a medical therapy very popular in the day and in that part of England in particular), it was a popular venue for lectures and a meeting place for learned men to discuss the fields of geology and palaeontology.

It did not take long for Grindrod's collection to come to the attention of a number of eminent geologists, and some of his specimens began to appear in their publications. Sir Roderick Murchison used Grindrod's collections in his book *Siluria*, 1859, and J.F. Blake illustrated a significant number in his monograph *British Fossil Cephalopods*, 1882. Grindrod was also an associate of the Oxford University Natural History Museum's first Keeper, Reader John Phillips.

The Grindrod Collection was bought by Oxford University Museum, now the Museum of Natural History, in 1883 for £500. The purchase was arranged by Joseph Prestwich, Professor of Geology at Oxford, and the price included the specimens, the cabinets and the cost of employing Robert Etheridge, then based at the British Museum (Natural History), to value the collection.

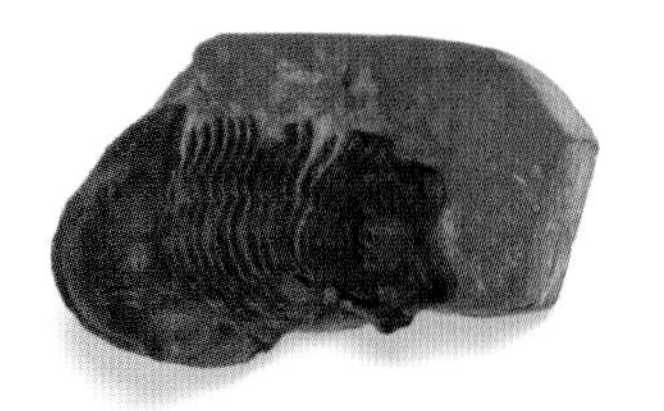

A letter from geologist Henry Woodward to Prestwich, dated 4 August 1885, provides further accolades for the Grindrod trilobites, now in their new home in Oxford:

> We have had two very fine days work at the Grindrod Trilobites in the Museum … I am putting the Wenlock Trilobites into fair order & writing labels & mounting some of the loose ones. I am also marking those which Salter has figured and which were without any indication to show which they were. … I will report to you later as to whether I need ask the loan of any of the Grindrod Trilobites to figure for the Palaeontographical Society. … The most beautiful specimens are of the genus *Deiphon* & *Sphaerexochus* – the former is, & always will be exceptionally rare.
>
> The result of my researches in the drawers of the Grindrod Cabinets is to confirm my previous good opinion of the valuable nature of the Collection which certainly delights me very much.

Dr Livingstone's tsetse flies

Three small flies, collected from the flanks of horses used by British traveller and explorer Captain Frank Vardon by his good friend Dr David Livingstone, were sent to London in 1850 for identification by John Obadiah Westwood.

He immediately recognized them as a new species and later that year described them as *Glossina morsitans*, and belonging to a group known as *sētsé* or tsetse flies. At the time little was known about these flies other than the fact that their bites caused a severe sickness in both livestock and people, often resulting in death. It would take over fifty years before the connection was made between them and the parasitic protozoa *Trypanosoma brucei*. In 1903, Major-General Sir David Bruce, a famous pathologist and microbiologist, recognized them as agent and vector for the protozoa that cause the disease better known as sleeping sickness.

The label accompanying the specimens bears an enigmatic reference to the 'Zimb of Bruce', in this case Bruce being the explorer James Bruce (1730–1794) and not the later microbiologist. There is a remarkable amount of early- to mid-nineteenth-century literature discussing exactly what this Bruce meant when he wrote of the Zimb in Africa, describing an insect, 'that much resembled a bee in appearance' and caused considerable distress in animals 'for when once attacked by this fly, their body, head and legs break out into large bosses, which swell, break and putrify, to their certain destruction' (Westwood quoting Bruce).

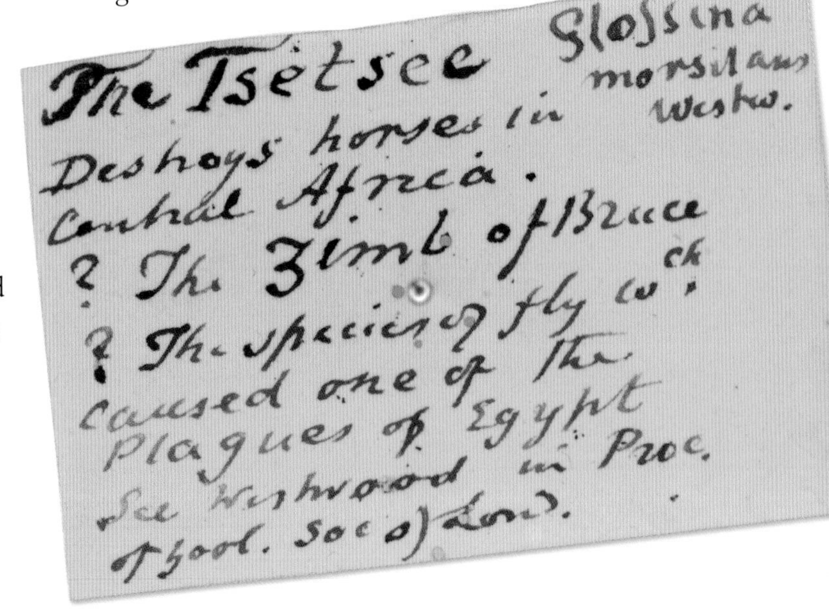

The name Zimb seems to have arisen from a phonetic translation of a local Aramaic dialect word for fly, and although it was hotly debated by numerous authors, it was adopted in the 1830s by entomologists and through publication of an annotated version of the Bible by D'Oyly and Mant in reference to the fourth plague of Egypt (that of an assortment of harmful insects, most typically depicted as a swarm of flies, which descended on the people). The label was clearly written early on in Westwood's research, presumably just after the specimens arrived in London.

The true nature of the Zimb seems finally to have been settled by Westwood. After comparing them with the tsetse flies and their habits he attributes the Zimb to the family Oestridae, or bot flies, the larvae of which are internal parasites in a variety of mammals.

Wallace's giant bee

Alfred Russel Wallace (1823–1913) is one of the least known, but most important, scientific figures of the nineteenth century. Considered by many to be the co-founder of evolutionary theory with Charles Darwin, Wallace studied the distribution of species in time and place, a science known as 'biogeography'. It was his work that finally prompted Darwin to publish *On the Origin of Species* in 1859, generating Darwin's reputation as the man who discovered evolution.

While the two men devoted their life's work to understanding evolution, their work and personal lives could not have been more different. Unlike Darwin, Wallace was a largely self-made man. Born into a large and financially unstable family (he was the eighth of nine children), Wallace was inspired by his love of the natural world. Later he wrote that he could recall virtually every detail of the environment of the River Usk near his childhood home, but could barely remember anything of his family from that time. His love of the outdoors led him to teach himself botany and entomology, and how to collect, identify and preserve animal and insect specimens.

In 1848 Wallace decided that he would travel abroad to follow his ambition of becoming a naturalist. Short of personal funds to support his expeditions, Wallace became a collector, bringing back specimens for sale in Britain. Collecting exotic and unusual specimens from around the world was an increasingly popular hobby of the wealthy and Wallace took advantage of the income it would bring. This would fund his academic work, giving him the time and reason to collect, observe and record his findings.

One of the most famous specimens that Wallace collected is now held in the museum. It is a single female bee, *Chalicodoma* (*Megachile*) *pluto*, discovered by Wallace in Indonesia between 1858 and 1859. Now known as Wallace's giant bee, it is the largest bee species in the world, believed to be extinct until 1981, when it was rediscovered by the American entomologist Adam C. Messer. The specimen has a 63 millimetre (2.5 inch) wingspan with which she would have completely dwarfed bees of other species. She also has enormous mouthparts, with jaws longer than her actual head, held agape in the same fashion as a stag beetle's mandibles. Unusual specimens such as these, with distinct characteristics and specific traits that suit their habitats and environments, led to Wallace's important discoveries in the Malay Archipelago about the process of evolution.

A letter from Charles Darwin

The theory of evolution is one of the best-known scientific theories and has made Charles Darwin one of the best-known scientists to have ever lived. Darwin did not work alone on his discoveries, however, and others were coming to the same conclusions at the same time, including Alfred Russel Wallace. Darwin was also an avid reader and acquainted with many scientists whose ideas paved the way for his publication of *On the Origin of Species* in 1859.

In addition to the work of other biologists, Darwin was heavily influenced by geological discoveries published over the past century, each emerging with increasing evidence that the formation of the Earth must have happened over a much longer period than biblical accounts suggested. His library contained many of these works and certainly supported his developing conclusions that life on Earth was not static, but that it changed and adapted over long periods of time.

One geologist with whom Darwin was in regular correspondence was the first Keeper of the Museum and Reader in Geology, John Phillips (1800–1874). Phillips was influential through his work on the development of the geological time scale, and the two wrote regularly to discuss his ideas. These letters to Phillips are now in the John Phillips Collection, held in the museum, and one such letter gives us an interesting insight into how Darwin anticipated the reception of his seminal work.

In this letter dated 11 November 1859, just under two weeks from the book's publication, Darwin writes to Phillips informing him that he will soon receive an abstract of his book. He asks Phillips to 'try not to condemn it utterly' and fears that he will 'be inclined to fulminate awful anathemas against it'. Darwin knew that despite growing evidence that the Earth was much older than the 6,000 years proposed by James Ussher in 1654, and even widespread acknowledgement in geological circles that the Earth was millions of years old, to propose that life on Earth was changing, and not a perfect creation of God, would be controversial, even heretical, to many at the time. Even scientists working on these theories struggled to understand the evidence presented to them outside a religious context. It would take many years, during which cultural and religious beliefs had to be overcome, before the theory was widely accepted as fact in all scientific circles.

does not require any answer
& pray believe me
Dear Phillips
Yours very sincerely
Charles Darwin

Down Brom.
Nov. 11th [1859]

My dear Phillips
I have directed Murray to send
you a copy of my book on the
Origin of Species, which as yet
is only an Abstract. — I fear
that you will be inclined
to fulminate awful anathemas
against it. I assure you that
it is the result of far more
labour, than is apparent in
its present highly condensed
state. — If you have time
to read it, let me beg
you to read it all straight
through, as otherwise it will be

Westwood's 'giant flea'

John Obadiah Westwood was born in Sheffield in 1805. He initially trained as a solicitor until it became obvious to all that he should pursue his real interest of natural history. In 1824 he met his soon-to-be patron Frederick William Hope, entomologist and later founder of the Hope Department of Entomology at the museum. Westwood devoted himself to the study of entomology, Crustacea and illustration, and in 1834 Hope appointed him to 'attend and arrange his insects'. As Westwood's reputation as an artist grew, he had opportunities to prepare illustrations for numerous natural history books, primarily of insects and Crustacea, and the museum's archives contain many of his original works. In 1861 he was elected the university's first Hope Professor of Zoology, a position he held until his death in 1893.

Westwood has long been recognized as one of the last great polymaths and was world-renowned for his expertise. Specialists and amateurs alike from around the globe would send him numerous specimens for identification, such was his standing in the natural history community. One fateful day in 1857, however, he received 'a most interesting specimen' that had been found by a Dr Backhouse of Gateshead. Westwood was fascinated by it, and so presented it the same year at a meeting of the Entomological Society of London, a society of which he had been a founding member in 1833.

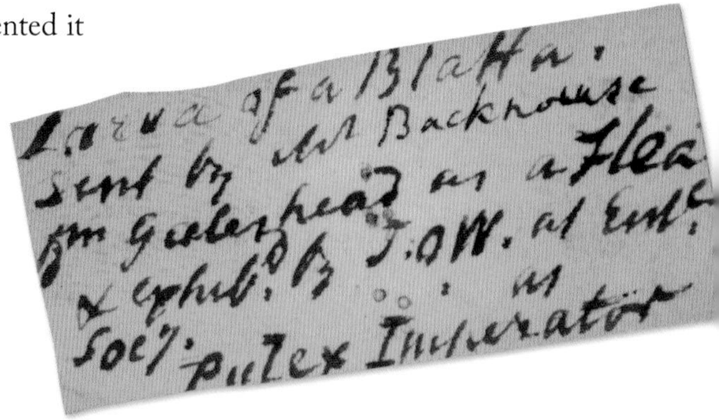

The specimen in question appeared to be a gigantic species of flea; Dr Backhouse discovered it in his bed and sent it to Westwood assuming it was a large example of *Pulex irritans*, or human flea. Believing it was a new species as it was twenty times larger than the common human flea, Westwood described it the following year, naming it *Pulex imperator* and implying its supremacy over all other described flea species.

Remembering that Westwood was a legendary entomologist, the next part of the tale demonstrates that even the most experienced scientist can make mistakes. In 1859, at another meeting of the Entomological Society of London, Westwood revealed that the specimen had proven not to be a flea at all, but instead was a very young and squashed larva of a cockroach, specifically *Blatta orientalis*, the oriental cockroach. The specimen still bears the name *Pulex imperator*, however, and is a reminder to all that even the best can get it wrong on occasion.

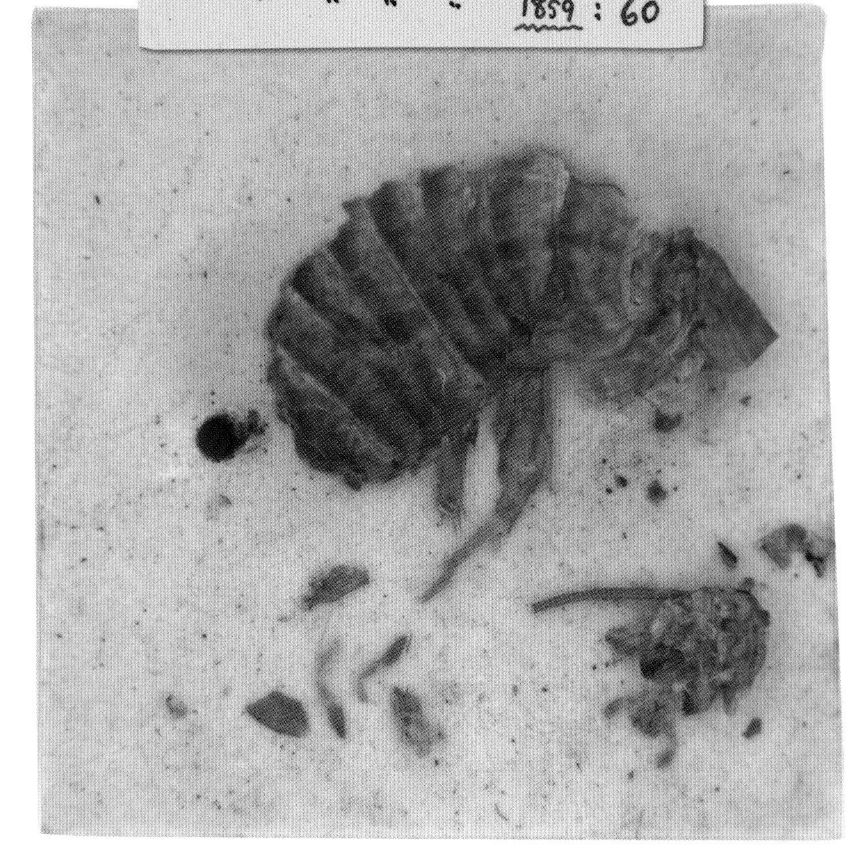

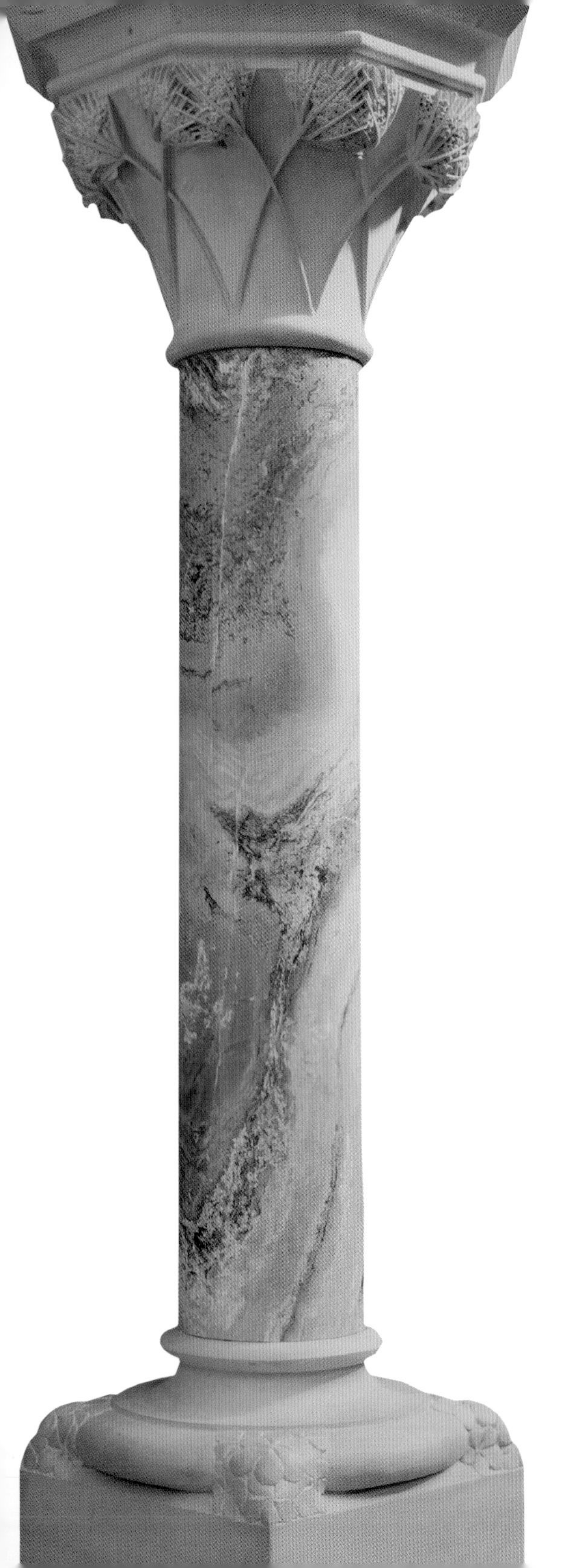

Pillars and carvings

The design and construction of the University Museum was very carefully planned. It was the intention of all those involved in its design and execution, both scientists and artists, that the museum would not only house the university's growing collection of scientific specimens, but also be a place of learning and research across the sciences. It was meant to inspire those who visited and nurture an appreciation of the natural world. This intention was woven into the design of the space.

The resulting building is a stunning and unique example of the Neo-Gothic architecture popular in Victorian Britain, in particular with its use of decorative patterns throughout. It was designed by Benjamin Woodward and Thomas Deane, but was also influenced by the design principles of John Ruskin (1819–1900), artist and critic, who, together with Henry Acland, was instrumental in bringing the construction and design of the museum to fruition.

Ruskin believed that artists must represent 'truth to nature' in their work, and inspired the group of artists that formed in 1848 as the Pre-Raphaelite Brotherhood. Many of the artists in the group would go on to contribute to the design of the museum, including designs for murals, fixtures and fittings and sculpture.

Some of the most striking examples of the building's architecture are the stone pillars, which show how the fabric of the museum, as well as the specimens housed within it, is used as a teaching tool. These thirty pillars, planned by John Phillips, first Keeper of the museum and Reader in Geology at Oxford, surround the main court, both on the ground level and along the upper gallery. Each is made of a different British decorative stone and is topped with a hand-carved corbel featuring a plant species taken from Oxford

University's Botanic Gardens. The columns are also flanked by carvings of plants from the same botanical order. It was initially planned that both the names of the stone and the plants would feature on the base of each column, but only the stone source and types were carved.

It was Phillips' intention to use examples of British rock, which traditionally might have been part of the museum's collections rather than its decoration, as a feature of the building's design. He selected several types of rock from various geological ages and locations to demonstrate their diversity, highlighting metamorphic and igneous rocks.

In addition to the capitals and corbels, there are also a number of statues in the front porch of the museum, surrounding several of the external windows on the front façade of the building, and along the underside of the upper gallery balustrade. The designs for the windows were made by John Ruskin and are now held by the Ashmolean Museum.

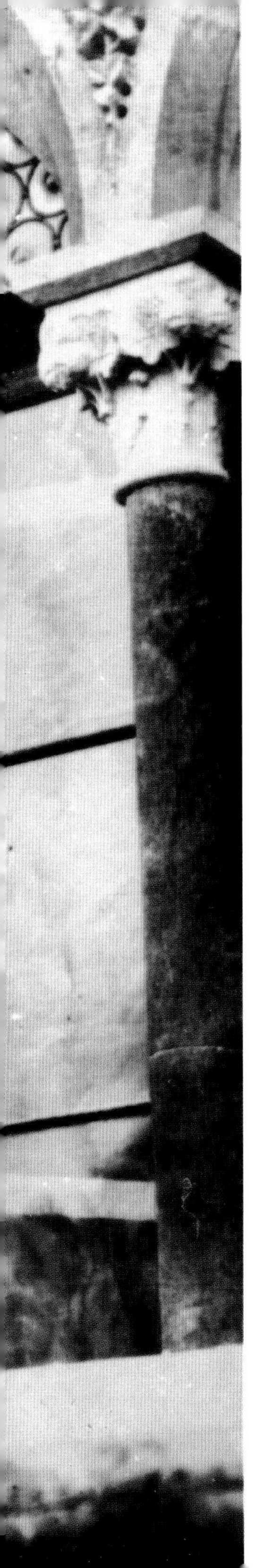

The majority of the carvings were done by a talented pair of brothers from Ireland, John and James O'Shea, and their apprentice and nephew Edward Whelan. The carvings were paid for through public subscription and the funds for forty-six carvings were raised between 1858 and 1860. Unfortunately, not all the carvings were done to the standard of the O'Sheas' work. Disputes over lack of payment meant they left the project before it was completed. The final corbels were carved in the early 1900s to an inferior standard and this contrast in the quality of carving can clearly be seen to this day.

Mer de Glace

The construction of the museum was a costly endeavour and the final total tallied in 1867 was triple the original estimate of £29,000. One of the most expensive features of the building was its decoration, and spiralling costs meant that the extensive and colourful murals that were to adorn the walls throughout the building were never completed, along with a number of other planned decorative features. The architect of the museum, Benjamin Woodward, had commissioned a number of murals to be designed by several of the famous Pre-Raphaelite Brotherhood and their contemporaries, but only those by a lesser-known associate, Richard St John Tyrwhitt (now better known as a writer), were finished.

Tyrwhitt (1827–1895) differed from the other Pre-Raphaelites in his use of photographs to complete his paintings rather than working *en plein air* (outside). *Mer de Glace* is an example of this difference in style and execution. Tyrwhitt had never been to the Mer de Glace in the French Alps and therefore the work was not based on his own knowledge of the location. Tyrwhitt first completed a study of the glacier on a board and then reworked this at a larger scale on the walls of the museum's Geology Lecture Room.

The subject matter was probably selected by art critic and friend to the Pre-Raphaelite Brotherhood, John Ruskin. Ruskin had visited the Mer de Glace a number of times in the years before the museum was built and had himself completed a number of studies of the rocks and other glaciers in the region. Ruskin, though best known as an art critic and artist, was also fascinated by geology.

This mural remains in the museum, along with another completed by Tyrwhitt. They are now in the office of the museum's director, at opposite ends of the room, along with the studies that inspired them.

Bell's tortoises

Originally trained as a dental surgeon, Thomas Bell (1792–1880) later followed his lifelong passion for natural history and, although not formally educated as a zoologist, he was highly successful in the field of natural sciences. In spite of 'merely' being an amateur researcher he was appointed Professor of Zoology at King's College, London, in 1836, and was a founder member of the Zoological Society of London. He was a member of the Linnaean Society, and chairman of the meeting in 1858 at which Darwin and Wallace jointly presented their ideas on natural selection, although by all accounts he was rather unimpressed.

Bell had a wide range of interests including amphibians, reptiles and crustaceans. He published a large number of books and scientific papers, including a description of the reptiles collected by Charles Darwin during his voyage on the *Beagle*. His *Monograph of the Testudinata* is considered to be an ambitious and pioneering work, aiming to encapsulate the world's living and extinct species of tortoises, terrapins and turtles. Unfortunately it was never completed as it ran into financial problems, relying as it did on individual subscriptions for each of the published parts. Eight parts were published, however, and of particular note are the exceptionally fine illustrations of specimens that accompanied the scientific descriptions. The forty hand-coloured plates were executed by James de Carle Sowerby and lithographically reproduced by Edward Lear, both highly regarded illustrators in their own right.

Bell often relied on others to lend him specimens for his work and only some of the material used found its way into his personal collection. This material was ultimately purchased in 1861 by

Frederick William Hope, major benefactor to Oxford University and founder of the Hope Department of Entomology. He presented the Bell Collection of reptiles and tortoises to the museum, although this was hotly refuted at the time by a number of other people who believed they were in possession of these important specimens. This elicited much debate in the literature until Bell himself published a letter stating that he had sold his large collection of Reptilia to Hope, including those specimens figured in his work.

Many years later, the unpublished parts of the monograph, including some plates by Sowerby, were purchased by publishers H. Sotheran Ltd, who decided to reissue the whole work. Bell declined to complete the text, so John Edward Gray, Keeper of Zoology at the British Museum, was asked to do it. In 1872, *Tortoises, Terrapins and Turtles Drawn from Life* by James de Carle Sowerby and Edward Lear was published.

Frederick William Hope

Frederick William Hope (1797–1862) was a wealthy curate, educated at Christ Church, Oxford. His career with the Church did not last long, however, and he devoted most of his life's work to the study of entomology (insects), inspired by the zoology lectures of Dr Kidd that he had attended as an Oxford student. Hope is best known for the generous gift of his collections and an endowment to Oxford University at the end of his life. This endowment saw the appointment of a professorship named for him in Zoology, Hope's close friend and colleague John Obadiah Westwood being the first holder, and the founding of the Hope Entomological Collections at the University Museum.

Hope and his wife, Ellen (née Meredith) were both avid and intellectual collectors of material related both to the arts and to natural history. During his life, and with both his and his wife's considerable wealth, they amassed an enormous collection of nearly 140,000 portraits and 90,000 engravings, including 20,000 items relating to natural history. Hope also had an enormous collection of entomological specimens from around the world. He and his wife recognized the importance of the collections they had accumulated and offered them to the university to be kept in its museums. The majority of the art and engravings went to the Ashmolean Museum, while all the natural history art and books, as well as the specimens, went to the new University Museum.

This oil portrait of Hope was presented to the university by his wife in 1864 shortly after his death. It was painted by Lowes Cato Dickinson, a student of Ruskin more famous for his engravings and lithographs than paintings. As was common practice at the time, the portrait was completed from a photograph belonging to the family. The portrait now hangs in the museum Reading Room, in what used to be known as the Hope Entomological Library.

Specimens in glass

These delicate models of British sea anemones were created by the Blaschkas, a family who specialized in glasswork and ran a business spanning 300 years and nine generations. But it was only from the late nineteenth century that Leopold Blaschka, a master lampworker and glass modeller, turned his skills to making models of microscopic organisms and soft-bodied invertebrates. He was later to be joined by his son Rudolf

Blaschka, and while they are most well known for their collections of glass flowers and plants, they did for a while have a thriving mail-order business specializing in marine invertebrates. Inspired by zoological specimens, scientific papers and observation of living animals, as well as artworks showing colours and structures that were difficult to preserve, the Blaschkas created thousands of glass models, many of which made their way into museum and university collections. The models' stunning colours and fine attention to detail made them valuable for teaching and exhibitions, as well as beautiful objects in their own right.

The models at the museum, acquired in 1867, are thought to be some of the oldest surviving Blaschka glass models. Even though they are over 150 years old, and in some cases slightly inaccurate representations of species, they still show the vibrant colours and alien shapes of British anemones in a way that is virtually impossible outside of anemones' living environments.

Darwin's portrait

This photograph is among the most iconic images of one of the best-known scientists to have ever lived: Charles Darwin. It was also his favourite. His declaration 'I like this photograph very much better than any other that has been taken of me' was printed at the bottom of many copies.

It was taken in 1868 in Freshwater, on the Isle of Wight, while Darwin and his family were on summer vacation. Though he looks older, Darwin was fifty-nine when he sat for the portrait, having only recently grown his famous beard. His career was well established and his theory of evolution had been published nearly a decade before. He was no stranger to sitting for portraits or receiving attention by this point in his life. The person behind the lens, however, is also noteworthy. The portrait was taken by early photographer Julia Margaret Cameron (1815–1878). She had only recently received her first camera at the age of forty-eight, just five years earlier.

Cameron's work revolutionized the medium of photography as a recognized art form. While she aimed for perfection in her photographs, as is demonstrated in her correspondence with contemporaries such as Sir Henry Cole, now housed at the Victoria and Albert Museum, she also experimented with the process. The style of her photographs played with focus and composition, giving her sitters an almost ethereal quality. She used the process of developing photographs to experiment with finish, unconventionally exhibiting works that others would have deemed failures, with smudges, scratches and even fingerprints appearing in the final works. Some have attributed her unrestrained enthusiasm for the process, learning along the way, to the fact that she took up photography later in life.

The copy of Darwin's portrait held in the museum's archive is most certainly an original, with Cameron's signature featuring on the back and her printseller Messrs Colnaghi's blind stamp on the front. It is also somewhat of a mystery, having only recently been rediscovered and with no record of how it was obtained by the museum.

Ch. Darwin

Antler flies and adaptation

The variety and diversity of form demonstrated among living plants and animals is truly breathtaking in its range. From the fleshy tentacles that make up a star-nosed mole's sensory receptor organ to the leafy protuberances of the sea dragon that camouflage it so well, these evolutionary adaptations can seem both bizarre and mysterious at first glance. Whilst some species have been well studied, many others have not, and there is still much to be discovered, especially among the invertebrates. Although over a million insect species have been described we know little about the habits and lives of many of them, and it is among the insects that we see a truly stunning array of adaptations.

The antler flies of the Bigot-Macquart Collection are a particularly fine example of this phenomenon. These specimens were collected in Papua New Guinea and belong to a subfamily within the Tephritidae, which are commonly known as antlered fruit flies. The antlers are a product of sexual selection and only the male flies have them. They are used to compete for females either through a direct comparison of size or through wrestling, during which males will lock antlers and attempt to push one another away. The antlers are actually enlarged cheek processes and are often coloured or patterned to aid in display.

The differences between the males and females are also an example of sexual dimorphism, where the different sexes have varying morphology or body forms. This difference may be highly pronounced as it is with the antler flies, or more subtle, such as with blackbirds, where the females

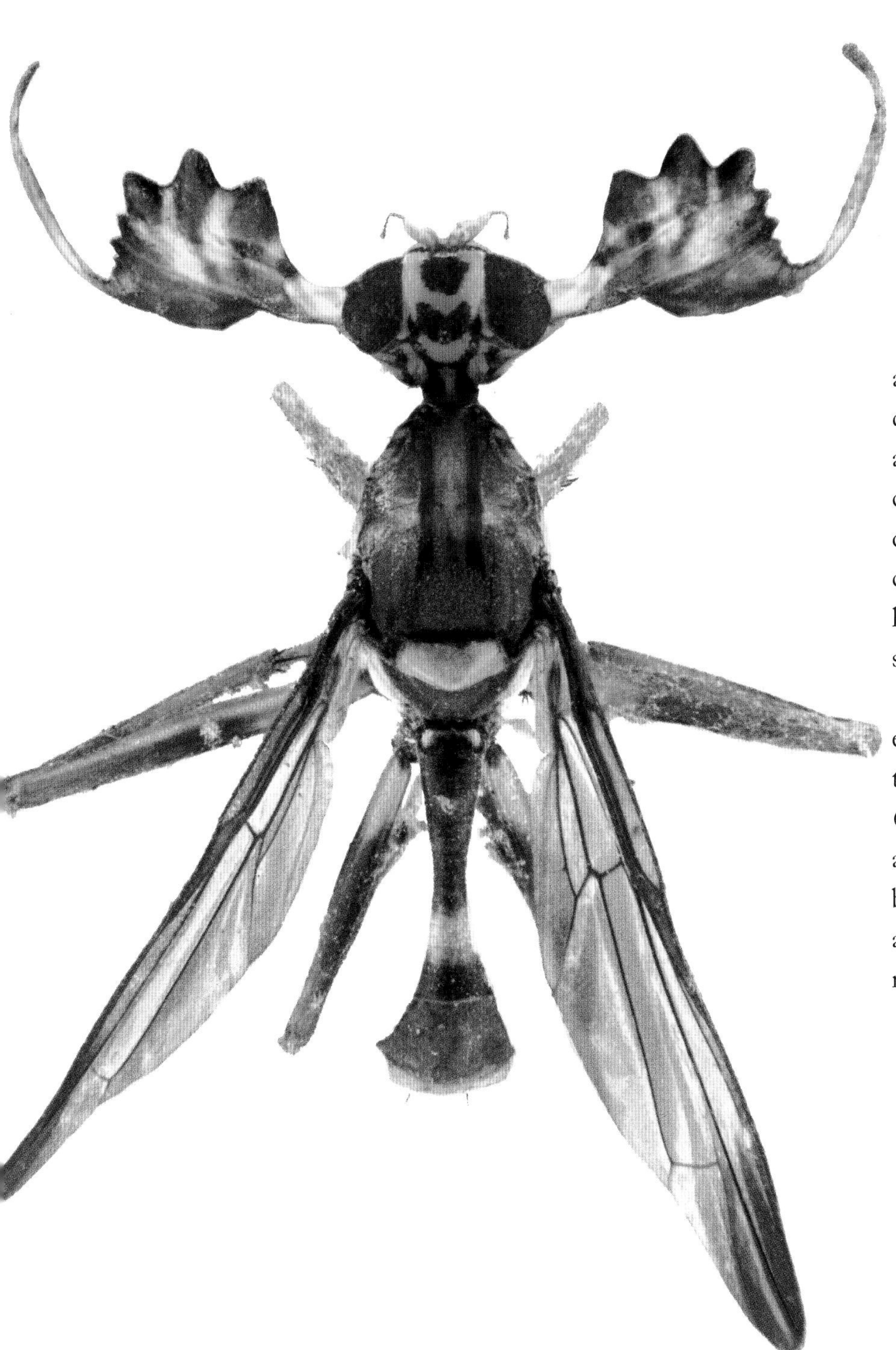

are a drab brown that helps camouflage them when they are incubating eggs or brooding chicks. In some instances the differences between the sexes can be so pronounced that they have been described as two separate species.

In another example of extreme sexual dimorphism, the females of the tussock moth *Orgyia recens* or Scarce Vapourer are wingless, living only for a brief period of time as an adult and relying on the winged males to locate them.

From entomology to anthropology

Museums preserve collections of material even when they are not needed for specific projects, because they never know when research will be able to throw new light on a subject or taxonomic group. It may take many years before the use for specimens is revealed, but careful preservation ensures they are there when needed.

One such example is the Denny Collection of Lice, which was obtained for the museum in 1871 upon the death of the original collector, Henry Denny of Leeds University.

Denny was the first curator of the Leeds Philosophical Society Museum and an avid collector of all types of animal specimens, but especially of lice of the world, going on to become an authority on the order Phthiraptera over the course of his life's work.

Lice are small, wingless insects that are parasitic on birds and mammals, living on the skin and feeding on the blood of their host. They are important disease vectors; the human body louse carries

infections such as typhus and trench fever, as louse infestation becomes more prevalent during times of war.

The Denny Collection comprises more than 3,000 lice as well as a document archive including an incomplete and unpublished book on exotic lice species following on from Denny's previous book, published in 1842, documenting the sucking lice of the British Isles. It even includes the original copper plates used to print the illustrations, which were then individually hand-painted by Denny himself. Specimens in the Denny Collection come from researchers the world over, including lice collected by Charles Darwin during the *Beagle* voyage, and from dozens of host species, from birds to seals to humans from Africa, Australia and South America.

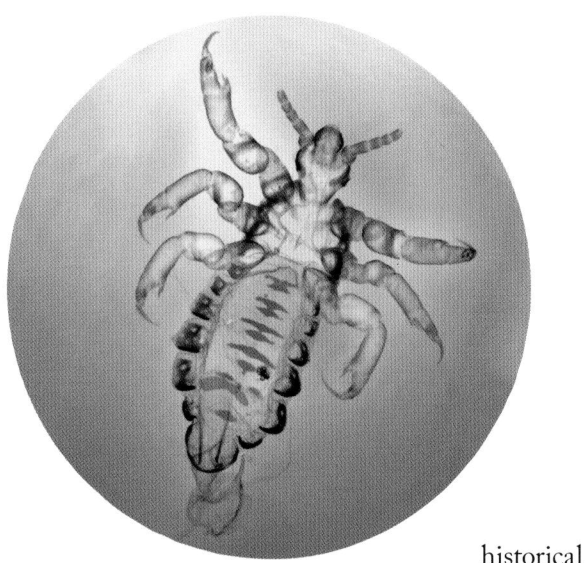

Many specimens are also from extinct species such as the passenger pigeon and thylacine or Tasmanian wolf. As lice are particular in their host preference, generally being exclusive to one particular species, the extinction of the host means that the louse also becomes extinct. The specimens that the museum holds are some of the last remaining evidence that these species ever existed.

Newly developed scientific techniques can be used to understand the currently tangled web of lice evolutionary history, and this knowledge is likely to expand to other historical collections and groups of organisms. Forensic scientists can sequence the DNA from the remains of the last blood meal held in a louse specimen's gut, allowing investigators to shed light even on cold cases using entomological specimens collected from victims. As the level of sophistication increases in science, the possibilities for study unfold, doubtless unimaginable to collectors such as Henry Denny.

The finest fossil fish

This beautiful example of a fossilized fish is *Dapedium punctatum* from the collection of the Philpot sisters, who lived in Lyme Regis at the turn of the nineteenth century (see p. 86). It is the holotype, the single specimen upon which the description and name of a new species is based, of an extinct Lower Jurassic (190 to 250 million years ago) species. It was a ray-finned fish with a near-circular body and short tail whose shape allowed it to make sudden changes in direction while swimming. Its pointed teeth and bony plated skull indicate that it was a predator, and there is evidence it ate hard-shelled invertebrates such as sea urchins and mussels, as it had a strong jaw designed to crush their shells.

In 1834, Jean Louis Rodolphe Agassiz (1807–1873), a Swiss geologist famous for his work on fossil fish, made his first visit to England and went to Lyme Regis to examine the Philpot Collection. He was mainly interested in examining fossil fish in connection with a publication he was working on and was fascinated by what he found. The Philpot Collection, now in Oxford, contains around 400 specimens from many different invertebrate and vertebrate groups, but is particularly rich in fossil fish.

The *Dapedium punctatum* specimen was included as plate 25a in Agassiz's *Recherches sur les poissons fossiles*, volume 2, part 1 (1835). The associated text, written in French, translates as:

The original of Plate 25 a, which represents a complete fish, is the finest example of fossil fish I have ever seen; it is also from Miss Philpot's collection.

On the death of the youngest sister, Elizabeth Philpot (1780–1857), the collection passed to a nephew, John Philpot, and in 1880 it was presented to the museum by his widow. Elizabeth Philpot is one of the best-known early female fossil hunters in British history, along with her contemporary Mary Anning.

Precious opal

This exquisite specimen is a rare example of a fossil made of precious opal, a type of gemstone. It is a gastropod, *Ampullospiro* sp., that would have lived in the sea or a river estuary in early Cretaceous times, around 110 million years ago. After it died, it was buried by sediments and chemical processes resulted in the original calcium carbonate of the shell being replaced by silica in the form of opal. The silica was likely derived when quartz grains in surrounding rocks were dissolved, although other processes, including microbial action, may have been involved.

Opal is an amorphous material composed of silica and variable amounts of water. In this case the silica is in the form of tiny spheres stacked up like a heap of ping pong balls, which can only be seen using a scanning electron microscope. Most opal found around the world has a waxy or glassy appearance and does not show 'opalescence'. Precious opal is a rare variety of opal in which the tiny silica spheres are just the right size and stacked in a very orderly way, so that when it is turned in the light it shows a play of colour; the colours seen will depend on the size of the spheres and the angle of the incident light. Most quality opal jewellery is made with this type.

This particular fossil was purchased by the museum sometime in the late nineteenth century. It comes from the opal mining fields in Australia, and was found in sedimentary rocks. This type of rock is formed through the deposition of sediment, particularly sediment transported by water, and its lithification (conversion to stone under pressure). It frequently contains fossil material.

Westwood's office

John Obadiah Westwood (1805–1893) is best known for his scientific contributions to entomology, the study of insects. Trained in law, Westwood abandoned the profession to pursue his love of the natural sciences and archaeology. He played a key role in the founding of the Entomological Society of London in 1833 and associated himself with some of the most eminent naturalists of the day. It was through these networks that he met the Reverend Frederick Hope, an amateur entomologist and wealthy patron to the science. Westwood worked extensively on Hope's enormous collection of insect specimens and oversaw Hope's donation of his one-of-a-kind collection to Oxford University in 1849. With this donation came an endowment for the Hope Professor of Zoology, and Westwood accepted the post, moving to Oxford from London to take up residence on Woodstock Road, not far from the museum – then newly built and the home to Hope's impressive collection of books, archives and insects.

In addition to Westwood's abilities as an entomologist, he was also a talented artist. He published pictorial works on a number of subjects, including early civilizations and religious theory, but it was his accurate and perfectly proportioned representations of insects that gave him the majority of

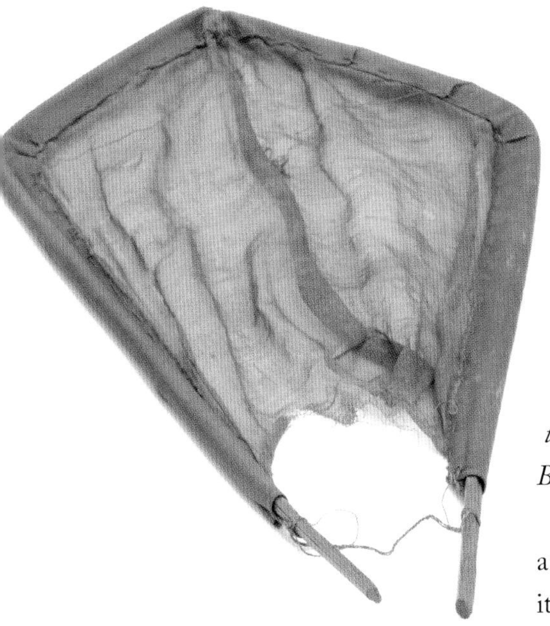

his commissioned work. The use of accurate scientific diagrams was a fairly new practice in the sciences in the nineteenth century, and was increasingly used in publications to provide to those without access to specimens the ability to see specific features. Westwood drew these illustrations for a number of entomologists, as most were unable to match his level of detail and accuracy. He also did the drawings for all of his own publications, including his most famous works, *An Introduction to the Modern Classification of Insects*, 1839, and *The Butterflies of Great Britain*, 1855.

As well as Westwood's enormous collection of papers, drawings and insects, the museum also holds some of his more personal items, including equipment which he used in his work and as teaching aides for his students. Most interesting is his clap-net, which was likely used to demonstrate the method of capture rather than for use in the field (field clap-nets are significantly larger, and would have been used to capture insects mid-flight). The collection also includes Westwood's spectacles, specimen pins and tools, his magnifying glass and even a pair of his shoes.

The Silver Bird Collection

Many of the collections held within the museum are historic in nature, having been donated in the nineteenth century or earlier. This material tells many stories, not just of these particular specimens but about the people who collected and donated the material, about those who looked after it once it was housed in the museum and about how that material has been used, whether it be for display, research or teaching. Sometimes those stories are exciting, full of bold adventures on field expeditions or new finds of fascinating species that change our perception of evolutionary history. On other occasions, however, the history of a collection may highlight the severe challenges of preserving organic material.

Steven William Silver was an avid collector with a long-lasting and wide-ranging interest in natural sciences and ethnology, and his collections were donated by his wife to both Oxford University Museum and the Pitt Rivers Museum after his death in 1906. During his lifetime he had amassed a wealth of material, including, as it was later described, 'a splendid and unique' collection of New Zealand birds. This set of specimens was put together for him by the foremost authority of the day, Sir Walter Buller, and included many examples of rare and scarce species, including those that are endemic to New Zealand and its surrounding islands. Since the collection was assembled many of the species have become extinct, including the featured bushwren, *Xenicus longipes*.

Unfortunately, between 1947 and 1950 it was found that many of the specimens were suffering from severe pest damage, one of the major factors in specimen destruction and one of the principal reasons that many historic items no longer exist in museum collections. Just over half of the collection had been destroyed beyond use or repair, despite staff efforts to save the material. Over fifty years later, damage to specimens remains an ongoing problem for museum staff around the world, and care and conservation of collections is always of primary concern.

Crescent nail-tail wallaby

The continent of Australia, isolated from other land masses and thought to be devoid of human inhabitants until as recently as 50,000 years ago, was of great interest to many explorers and specimen collectors. From an ecological perspective it has distinct and unique fauna and flora with numerous endemic species. Of particular fascination to Western scientists was the discovery of marsupials, a distinct group of mammals that give birth to their young when they are relatively undeveloped, which they then carry in a pouch. Many placental mammal species found in other areas of the world have marsupial counterparts, which evolved to fill equivalent ecological niches to species such as rabbits and small deer in a striking example of convergent evolution. The ecosystem functions of the rabbits and deer are instead fulfilled by animals such as kangaroos and wallabies.

The influx of new settlers to Australia in the eighteenth century was accompanied by an influx of non-native animals. Livestock such as sheep, cows, pigs and rabbits were transported as well as those with perhaps less obvious uses, such as cats and foxes, introduced as pets and for recreational hunting respectively. This has had a devastating effect on the native fauna. Some of those early specimens, collected in the early 1900s, such as the crescent nail-tail wallabies illustrated here, are now the only records of species that once inhabited Australia. The

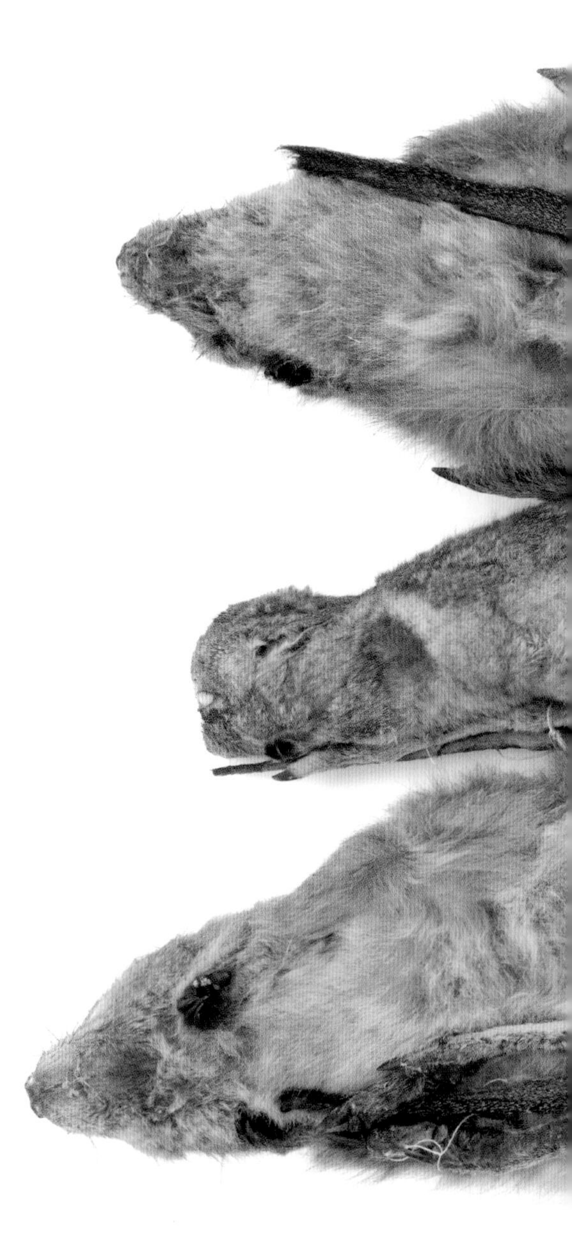

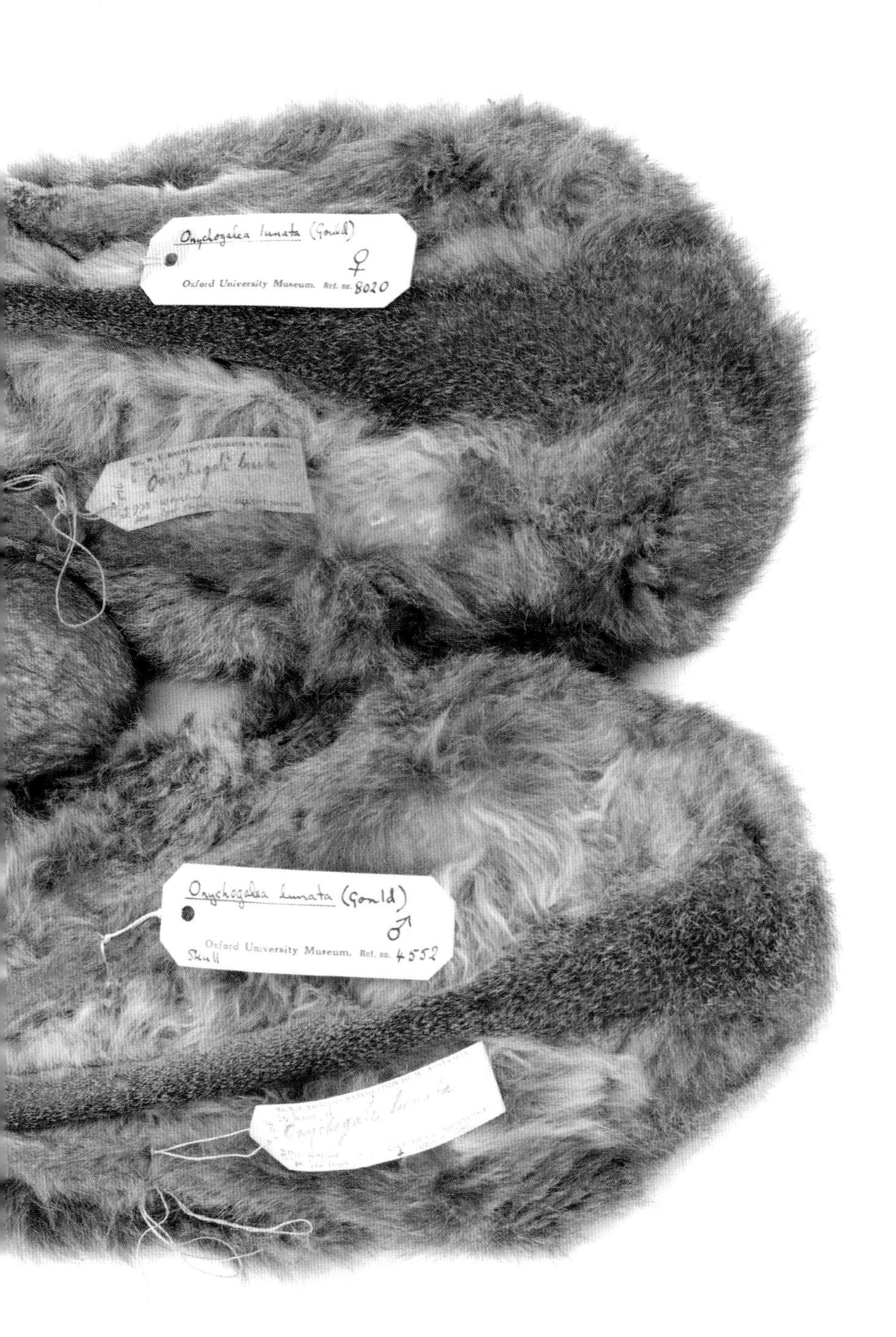

nail in the tail refers to the small horny spur that these wallabies possessed on the end of their tails. Scientists have as yet been unable to determine whether this spur served any purpose. Relatively abundant, though limited in their distribution, at the time that the settlers arrived, the crescent nail-tail wallaby was extinct by the 1950s. Known to be in decline from 1900 onwards, it is thought that the red fox ultimately brought about its extinction.

Aye-aye specimen

This soft-tissue specimen is one of two aye-ayes at the museum and was collected in 1911 thanks to a donation of £100 that same year. With this the museum was able to secure the services of the zoologist Paul Ayshford Methuen, who travelled to Madagascar with renowned herpetologist John Hewitt to collect and describe scientific specimens. Their remit was to collect the more important fauna of the island, the aye-aye included.

Daubentonia madagascariensis, commonly known as the aye-aye, is the world's largest nocturnal primate and is native to Madagascar. Compared to diurnal (daylight) primates, the aye-aye is relatively small, with adults having a body length of 360–440 millimetres and weighing just 2–3 kilograms. Despite its small size, however, it has a big reputation, being viewed as the harbinger of death by the native peoples. This is due to its extreme physical adaptations, which make it a strange and ghoulish-looking animal to many.

These adaptations include large dark eyes and a long middle finger, which, according to Malagasy legend, is used to pierce the hearts of sleeping humans. Its long digit is, in reality, used for locating insect grubs living under the bark of trees. The aye-aye taps its skeletally thin middle finger on the branch or trunk of a tree and listens to the reverberations through the wood with its large sensitive ears. Once a grub has been located, it uses its chisel-like teeth to excavate a hole to retrieve the meal. The teeth of an aye-aye are like rodents' teeth; they grow continuously throughout their lives and are incredibly strong, able to break into coconuts and other fruits and seeds.

The aye-aye's ghoulish appearance has been one of the reasons for its decline. Perceived as an omen of death, these animals were once

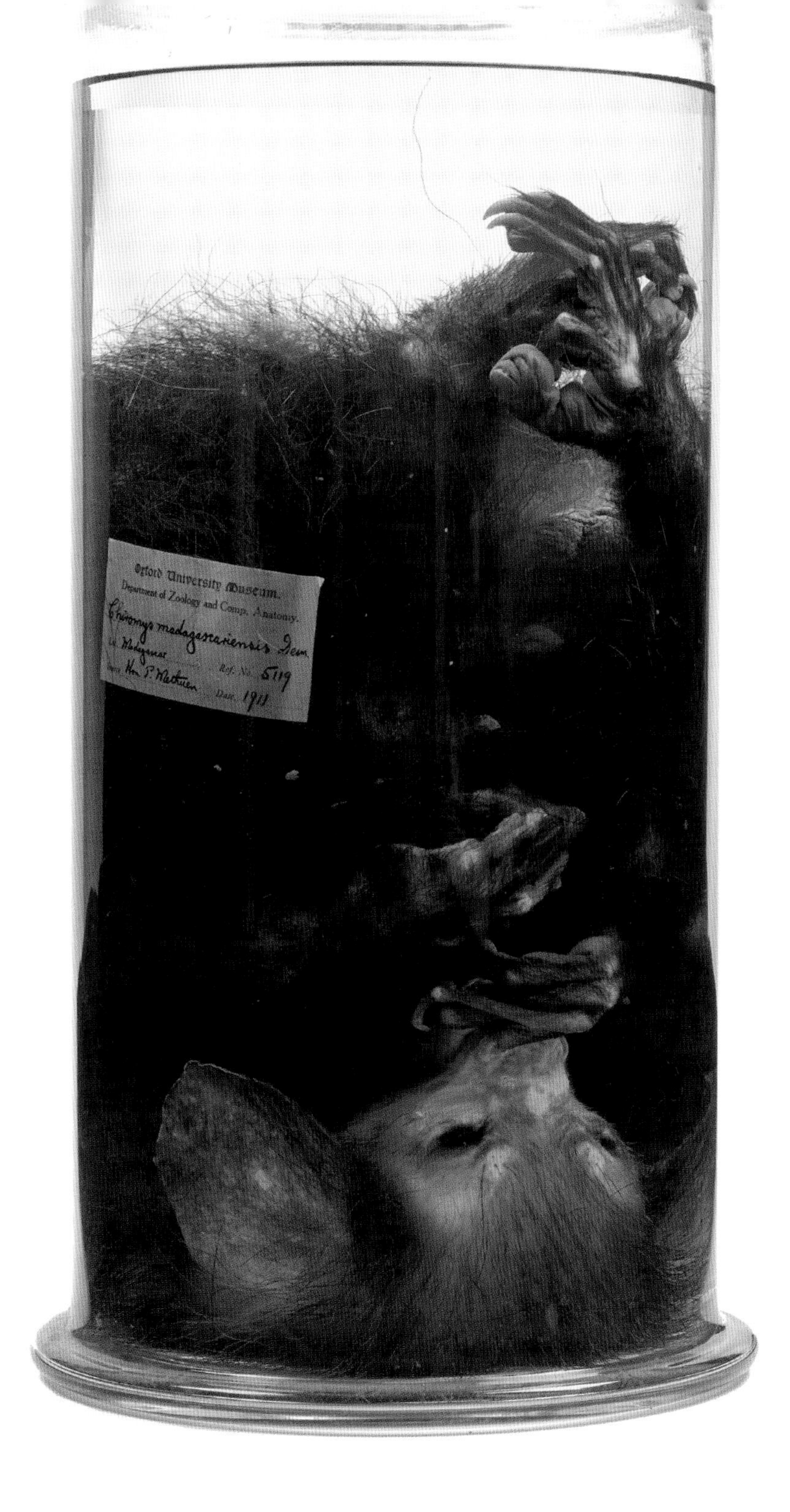

killed on sight. However, the main reason for their current decline is habitat loss, which has led to the aye-aye being listed in Appendix I of the Convention on International Trade in Endangered Species of Wild Fauna and Flora (CITES) and classed as 'Endangered' on the 2006 International Union for Conservation of Nature (IUCN) Red List of Threatened Species.

There is one other aye-aye specimen in the museum collections, a mounted skeleton which can be seen hanging from a branch in one of the museum's primate displays, revealing its long middle finger.

Typical Flies

Three volumes entitled *Typical Flies* were published between 1915 to 1928. Designed to draw people's attention to flies and to increase interest in the group, they are one of the earliest photographic accounts of flies and represent one of the first attempts at creating a field guide intended for use by the general public. Authored by Ethel Katharine Pearce (1856–1940), the volumes illustrate typical examples of species from the majority of major fly groups found in the British Isles.

The production of these volumes must have been something of a labour of love. The photographic equipment that Pearce used was either second-hand or homemade, and it took almost a day to set up, photograph and then develop an image. The specimens were specially mounted, with bodies, wings and legs presented on as flat a plane as possible. Even so, Pearce had to develop a technique for manually stopping down, that is, decreasing the size of the camera aperture in order to increase the depth of field and keep as much of the subject as possible in focus.

The specimens were mostly collected from the areas near to her home, which at the time yielded an abundance of rare species due to the mosaic of habitats that cover much of

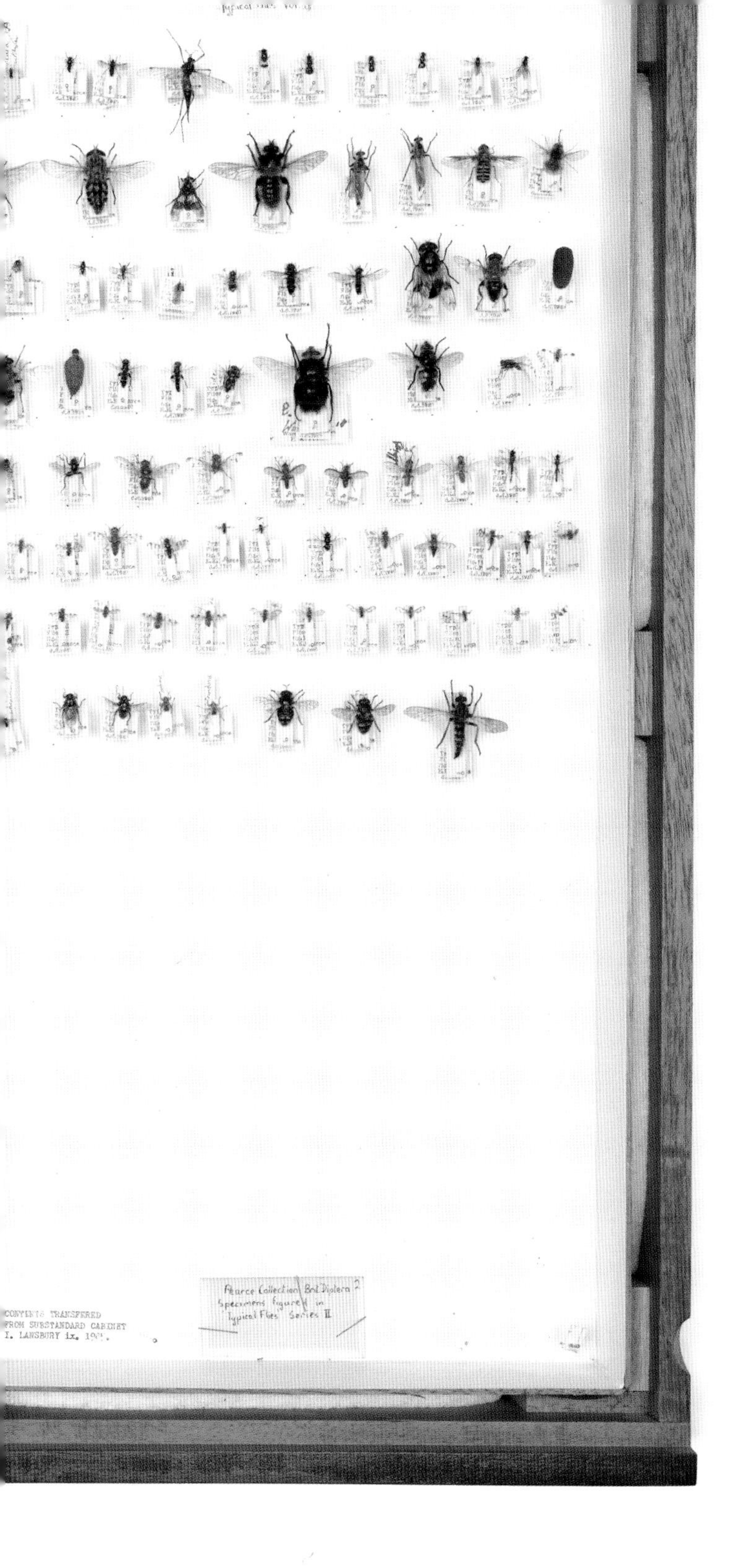

Dorset. Notes on capturing and preparing specimens for identification were included in each book, along with some photographs of the habitats in which species might be found. Many of the research entomologists of her day were full of praise for Pearce's work, contributing specimens and corresponding with her on the behaviours and habits of the species.

An illness during her school years reportedly left Pearce at least partially deaf, so it is interesting to discover that, when not occupied with photographing flies, she developed a career as a journalist, reporting on local news stories in her home parish of Morden in Dorset. She wrote for numerous publications, including a variety of newspapers and magazines, as well as producing a small number of scientific publications.

The collection of specimens that she used for *Typical Flies* is housed in the museum, marked by handwritten labels to denote the plate number in the volume in which they appeared. They remain as a discrete collection, arranged as they are in Pearce's publications, as a testament to one of the first female dipterists.

Goodenough swallowtail

This is one of the world's rarest butterflies on account of its highly restricted range, only known from a few high peaks on Goodenough, a small island in the D'Entrecasteaux group in the Solomon Sea. The Goodenough swallowtail is a subspecies of *Graphium weiskei*, known as the purple spotted swallowtail, which is endemic to the nearby island of New Guinea. Over time, the colouration of the Goodenough swallowtail has diverged sufficiently from the mainland species that it is identifiable in its own right.

This is the first specimen of this butterfly ever captured. It was collected by the anthropologist Diamond Jenness during a year-long research stay on the island between 1911 and 1912. A close family tie had afforded him the opportunity to stay there, where he studied the native society and collected cultural objects that were to be donated to the Pitt Rivers Museum.

While studying at Oxford on a scholarship, Jenness became acquainted with Edward Poulton, then Professor of Zoology at the University Museum, and agreed to collect specimens for the museum. During a trip into the mountainous interior of Goodenough, Jenness encountered this specimen, but lacking a net or any other equipment, was forced to improvise and struck the butterfly with a tree branch (resulting in damage to one of its wings).

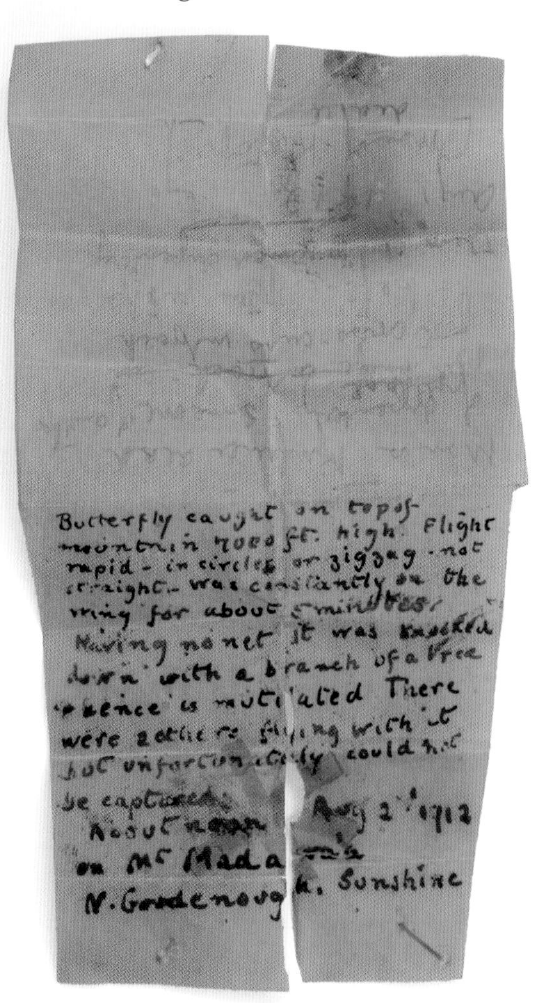

Butterfly caught on top of
mountain 7000 ft. high. Flight
rapid - in circles or zigzag - not
straight - was constantly on the
wing for about 5 minutes.
Having no net it was knocked
down with a branch of a tree
hence is mutilated. There
were 2 others flying with it
but unfortunately could not
be captured. Aug 2nd 1912
About noon
on Mt Mada
N. Goodenough. Sunshine

Described as *Graphium weiskei goodenovii* in 1916, for many years this butterfly was only known from this and another specimen obtained a year later by one of Walter Rothschild's collectors. Recent expeditions to Goodenough have sighted this butterfly again, still flying in its remote mountain habitat.

Inflated caterpillars

Inflating is a highly specialized technique that is used to preserve caterpillars in a life-like condition without shrivelling. The alternative for those studying caterpillars is to preserve specimens in alcohol but this is less than ideal since characteristics needed for identification are often lost or obscured during the preservation process. To obtain near-perfect specimens through inflation, however, like the ones shown here, takes much skill and practice and is something of a lost art.

The apparatus used to blow caterpillars consists of two parts: lengths of fine glass tubing and a glass flask, held horizontally on a stand and heated by a small alcohol burner.

First, an incision is made at the rear of the caterpillar. Starting at the head end, the internal organs of the caterpillar are squeezed out, in the manner of squeezing toothpaste from a tube. The glass tubing is then inserted into the rear incision and the skin tied off to create an airtight seal. The preparator blows very gently down the tubing, inflating the skin in much the same way as one does a balloon.

To dry the skin, it is held in the hot air inside the flask, all the time maintaining just enough pressure to ensure the caterpillar remains inflated at exactly the right size. Once dry, the caterpillar skin is attached to a pin, usually by gluing it to a length of wire or wooden splint. In some instances the loss of internal organs and fluid layers means that what were once vibrant colours on the living organism are lost. Some collectors took great pains to recreate these by hand painting the dried skins using watercolours.

Omnium 46
Insectorum
Blown larvae
T. A. Chapman

The Colours of Animals

Edward Bagnall Poulton (1856–1943) was the second Hope Professor of the entomology collections of the museum between 1893 and 1933. Over the course of his tenure he made an enormous contribution to the museum's collection, adding many hundreds of thousands of specimens. Such was his zeal in acquiring specimens that he earned the nickname 'Bag-all'.

Poulton was particularly interested in mimicry, a fascinating process whereby one species evolves a resemblance to another, often to gain protection from predators. His book *The Colours of Animals*, published in 1890, was the first comprehensive guide to the function of colours in the different animal groups, such as mammals, birds and especially insects.

In direct contrast to the entomology department's first Hope Professor, Poulton was a keen proponent of the theory of evolution. Convinced that mimicry was the result of natural selection, he published many new and remarkable examples of the phenomenon in insects to support his

ideas. One particular example is given here, where a swallowtail butterfly *Papilio laglaizei* has evolved a striking resemblance to a distasteful day-flying uraniid moth *Alcides aurora*.

These two specimens were captured flying together in the rainforest of south-east New Guinea and given to the museum by the famous zoologist and collector Lord Walter Rothschild. In a paper published in 1931 Poulton concluded that when the butterfly folds its wings the two orange hindwing spots are brought together to appear as if they are the bright orange abdomen of the moth.

Papilio laglaizei ♂, and model!

Tr. Ent. Soc.
79. 1931
Pl. XIV. figs. 1, 2

Wager's rocks

One of the most important collections of rocks held by the museum is that donated by Lawrence Rickard Wager (1904–1965), Professor of Geology at Oxford, mountaineer, explorer and one of the greatest geological thinkers of his time.

Wager is best known for his work in the Skaergaard area of Greenland. He discovered the Skaergaard Intrusion, a layered igneous rock formation, during Gino Watkins' British Arctic Air Route Expedition in 1930. A rock intrusion is formed when magma crystallizes and solidifies. A description of the intrusion was published in the journal *Meddelelser om Grønland* in 1939, and immediately became a seminal work in petrological research as the first detailed study of a layered basic intrusion. Its discovery led to several other important developments in petrology, the study of rocks.

Wager was also well known for his attempt to climb Mount Everest in 1933, returning with many petrological and mineralogical specimens. On 29 May, along with P. Wyn Harris, he managed to reach 28,100 feet, less than 1,000 feet short of the summit. The museum has a fascinating collection of 244 rocks collected by Wager on this expedition. Pictured here is 'Specimen 124 Grey metamorphosed limestone from the First Step at 27,890 feet', measuring approximately 3 centimetres across. This rock forms much of the topmost part of the mountain, around the summit.

Wager's collection also includes an extensive archive of papers and photographs that features hundreds of images of his many expeditions, including Skaergaard and Everest, making it both scientifically and historically important.

MOUNT EVEREST EXPEDITION
1933
GOGGLES
MADE BY
THEODORE HAMBLIN, LTD.
DISPENSING OPTICIANS
15, WIGMORE STREET, LOND

A rediscovered coelacanth

A coelacanth was thought to be a kind of fossil fish, long extinct, until its dramatic rediscovery in 1938 by Marjorie Courtenay-Latimer, curator of a small museum in South Africa. Prior to this, the coelacanth was only known from fossilized remains and was thought to have died out 66 million years ago. It was therefore something of a shock when a single specimen was landed at the docks of the port of East London on the Eastern Cape, coming ashore on a trawler that had been working on the east coast of Africa. The captain of the ship telephoned Courtenay-Latimer, as she often examined his catch in case there was anything suitable for the museum. She later described the moment she saw the coelacanth: 'I picked away the layers of slime to reveal the most beautiful fish I had ever seen', she recounts. 'It was five foot long, a pale, mauvy blue with faint flecks of whiteish spots; it had an iridescent silver-green sheen all over…' (Weinberg, *A Fish Caught in Time*).

Coelacanths belong to a group known as lobe-finned fish, which are more closely related to species of lungfish, reptiles and mammals than they are to other fish. They are large creatures that can grow to over 2 metres in length and weigh in the region of 90 kilograms. They have a number of unique adaptations, mostly notably

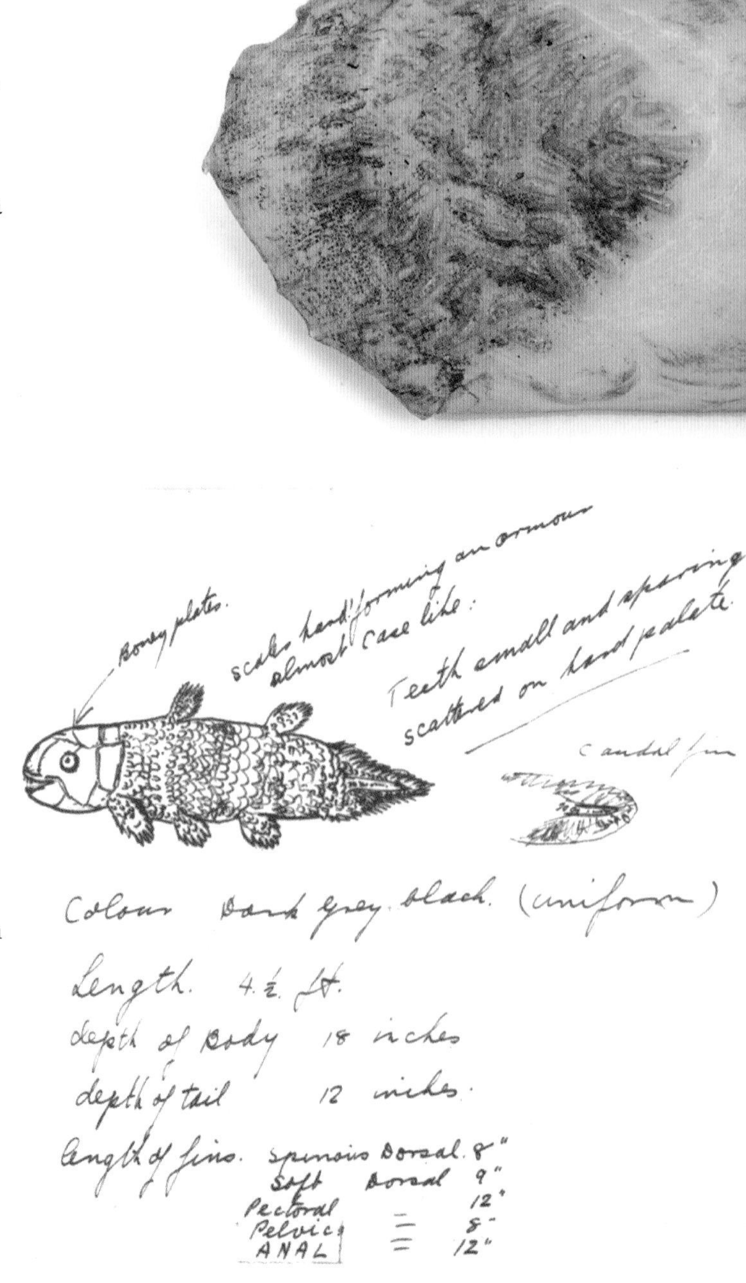

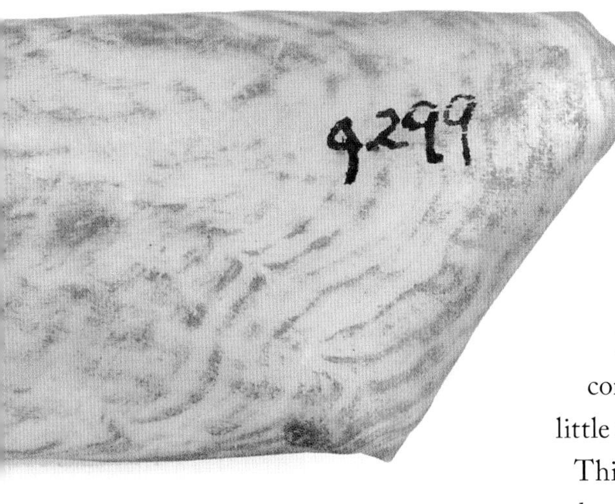

a fat-filled lung that functions in much the same way as other fish's swim bladders, being used to control buoyancy. They are often referred to as 'relict species', meaning that they have not evolved since the form seen in the fossil record. However, it is wrong to believe that coelacanths have not evolved, though it is true that they have done so very slowly, as recent DNA work has revealed. The combination of stable environmental conditions and lack of predators has meant that there has been little in the way of evolutionary pressure to force change.

This particular specimen, landed in 1938, was difficult to move and proved to be impossible to properly preserve in a small town that had extremely limited supplies of formalin and no freezer. Desperate attempts were made to keep the specimen as intact as possible while Courtenay-Latimer sent a description and sketch to local expert, Professor James Leonard Brierley Smith, requesting his help to confirm the identity of this strange but beautiful fish. Decomposition forced the situation, but with the help of a local taxidermist the skin and skeleton were retained, and when Smith finally arrived in East London over a month later, the taxidermy mount was identifiable as a coelacanth and was described and given the Latin name of *Latimeria* in honour of its discoverer.

The museum has only a few small fragments, in the form of scales, from this original specimen, and whilst they may look like nothing much at all, they are linked to one of the most dramatic rediscovery stories in recent times.

Conversation with Smaug

Christmas lectures are a tradition in the museum. As one would expect, they are usually on a variety of topics related to natural history and ordinarily geared at a family audience. On 1 January 1938 John Ronald Reuel (J.R.R.) Tolkien (1892–1973) delivered a Christmas lecture on the subject of dragons. While this might sound like an unusual topic for a natural history museum, it seems less so when considering both Tolkien's work and the mythology of long-extinct species and their fossilized remains.

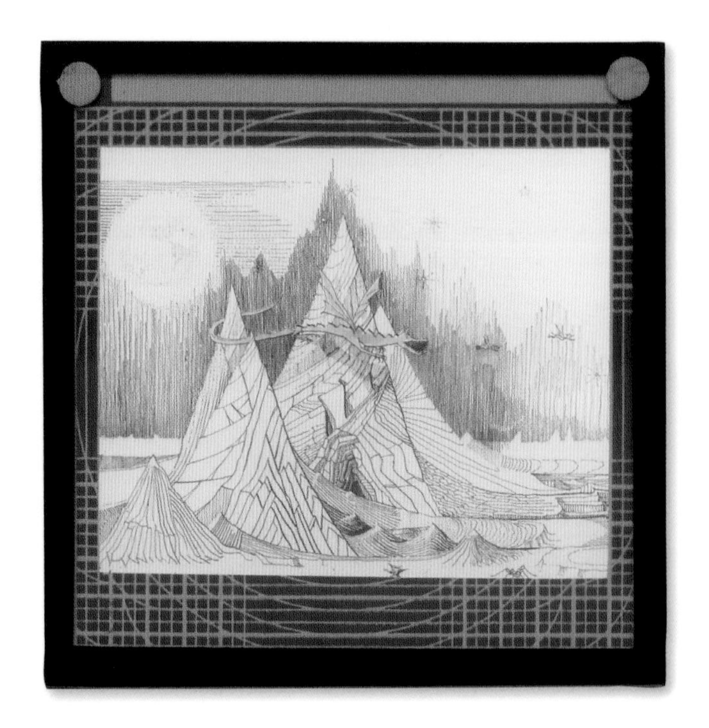

Tolkien's lecture was primarily on the mythology of dragons, but his slides also included a number of dinosaurs. He alluded to the similarities between the two types of creature, writing in his notes that 'it is an odd thing that scientists who are not concerned with legends make pictures strangely similar to terrible monsters'. He also discussed some possible inspirations for dragons from extant species, including crocodiles, snakes and lizards, and cultural links between these and European and Chinese dragons.

Interestingly, Tolkien's papers, now held in the Bodleian Library, reveal that he did not feel the lecture was well received, noting that perhaps it had not been made to the right crowd and also that he had not been feeling well. At that time he could not have anticipated the huge success that *The Hobbit*, published only three months before, would become.

This piece of popular culture history was forgotten until a few years ago when a visitor to the museum's entomology department discovered a box of glass lantern slides inside a large cupboard. As a fan of Tolkien, and aware that Tolkien had given the lecture just after the publication of *The Hobbit*, he immediately recognized the author's distinctive illustrations. It was an enormous surprise to both him and the staff, and, after some research, the story of the lecture quickly emerged. The slides, which depict drawings of 'The White Dragon pursues Roverandom & the Moondog' (left), 'Conversation with Smaug' (below) and a coiled dragon (right) are now housed in the museum archive.

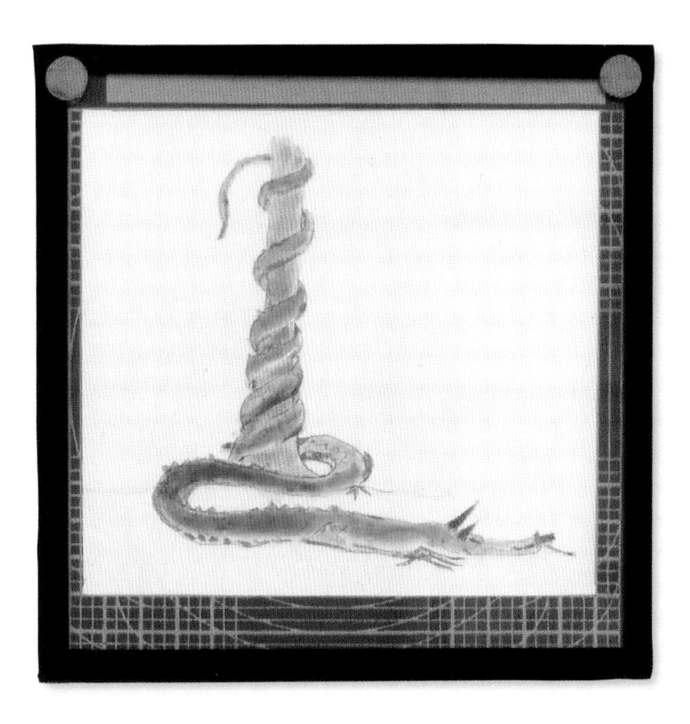

Boxer mantis

These two specimens are nymphs of the boxer mantis (*Oxypilus* sp.). They were collected by Dr Eric Burtt in 1942 during his travels in East Africa. Born in 1908, Burtt was fascinated with collecting insects from a young age; by the age of fourteen he had found all the British species of bumble bee. Burtt studied entomology at Imperial College, London, which led to an offer of a research scholarship at the Colonial Office. The position would have fulfilled his ambition to study in the tropics; however, a motorcycle accident, which caused permanent paralysis of his left arm, ended this career.

Determined to follow through on his plans, in 1934 Burtt travelled to Tanganyika (now part of Tanzania) and found work with the Medical Department on tsetse fly research projects. In 1939, he became a junior entomologist in the Tsetse Research Department, and it was during one of his trips to Tinde, within Tanganyika territory, that he spotted these two mantises, which were camouflaged on a gravel path. Being an avid collector, he sent the specimens and his written observations on their behaviour to Professor G.D.H. Carpenter, Hope Professor of Zoology, explaining what he had witnessed. This account was subsequently published as an article in the *Proceedings of the Royal Entomological Society of London* in 1945, in which Burtt began by describing how 'the Mantis needs to be observed alive for its interesting features to be duly appreciated, and to do it justice records should be made with a cine-camera'.

These nymphs were the models for Burtt's remarkable drawings, which illustrate the account he wrote of a threat display he saw one of the nymphs perform. He explained that the mantis held itself up almost vertically, with its antennae moving rapidly and one front leg stretched out forwards and downwards

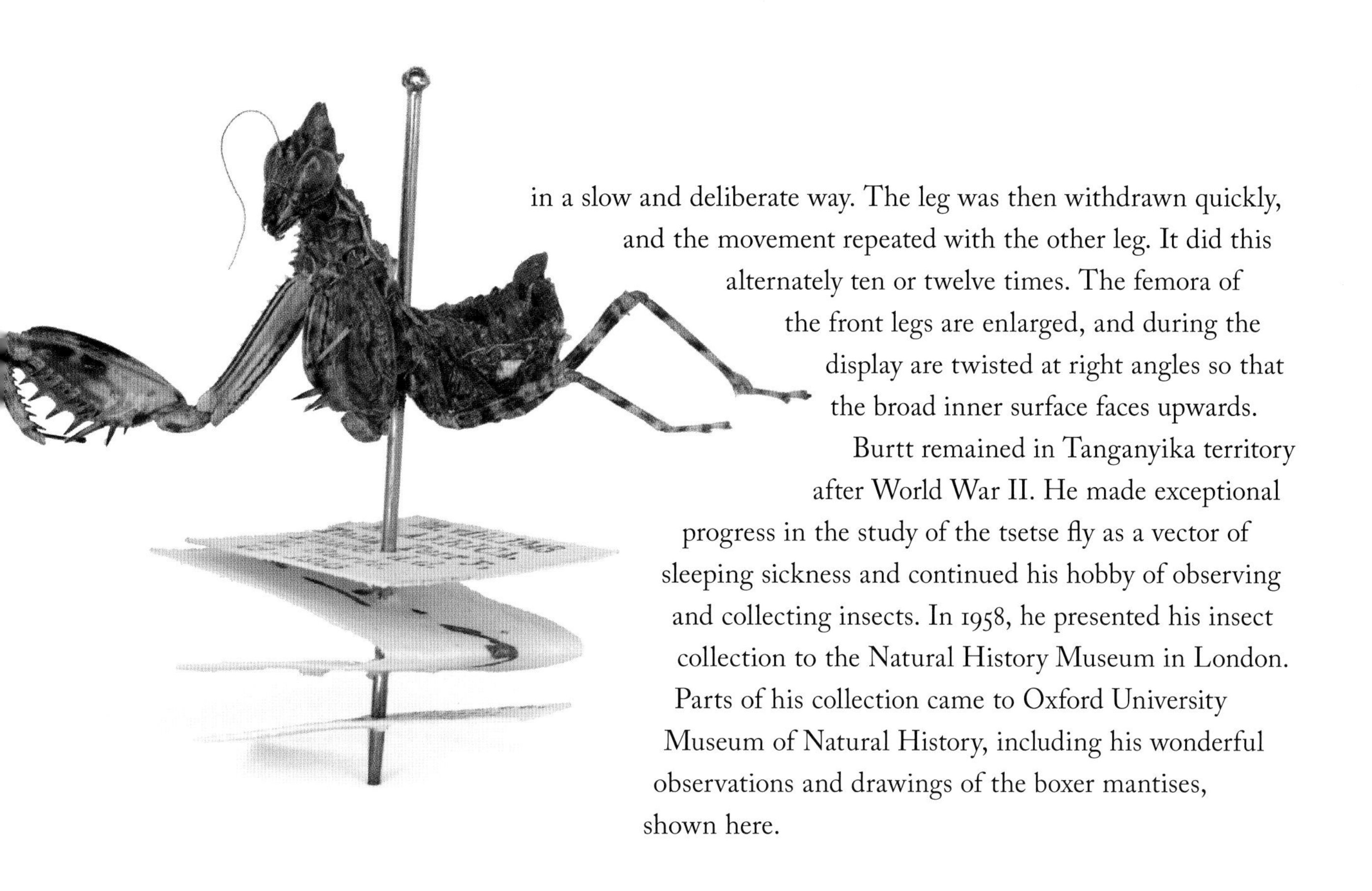

in a slow and deliberate way. The leg was then withdrawn quickly, and the movement repeated with the other leg. It did this alternately ten or twelve times. The femora of the front legs are enlarged, and during the display are twisted at right angles so that the broad inner surface faces upwards.

Burtt remained in Tanganyika territory after World War II. He made exceptional progress in the study of the tsetse fly as a vector of sleeping sickness and continued his hobby of observing and collecting insects. In 1958, he presented his insect collection to the Natural History Museum in London. Parts of his collection came to Oxford University Museum of Natural History, including his wonderful observations and drawings of the boxer mantises, shown here.

Swifts in the tower

Swifts had been nesting inside the ventilator shafts of the museum tower for many years when David Lack, Head of the Edward Grey Institute at the Department of Zoology, began the Swift Research Project. Swifts had proved a difficult species to study as they spend most of their lives in the air and choose nesting sites which are inaccessible to predators to safeguard the eggs and chicks.

The swift colony in the museum proved to be ideal for long-term research and the colony has been studied since May 1948. It is one of the longest continuous studies of a single bird species in the world.

Swifts are migratory birds and spend almost their entire lives on the wing. European swifts from Britain winter in Zaire, Tanzania or Zimbabwe. They have long, thin wings for efficient gliding flight over long distances. Their beaks have a wide gape to catch a variety of insects and tiny spiders while flying. They land only to breed and can fly at least 560 miles a day, gathering food, during the breeding season. They nest almost exclusively in urban areas and are extremely faithful to their nesting site, returning to places such as the museum tower year after year. But they are also extremely sensitive to change, and will abandon nest sites if disturbed. It is thought that urban redevelopment may be one of the reasons that numbers in the UK have fallen by 38 per cent since 1994. Consequently, the birds face an uncertain future.

In August 2016 the museum was delighted to announce its part in the partnership project 'Swift City', which was awarded a Heritage Lottery Fund grant to undertake vital research into the decline in the swift population. Run in conjunction with the Royal Society for the Protection of Birds (RSPB) and a variety of other organizations, the two-year project uses volunteers to map every swift nesting site within the Oxford ring road and add this knowledge to the council and university estates planning documents so as to maintain current swift nesting sites in the city, and create 300 further sites on new and existing buildings, in an effort to conserve this enigmatic species.

Dorothy Crowfoot Hodgkin

Many British men have received a prestigious Nobel Prize for their contributions to science, but only one British woman has been so honoured: Dorothy Crowfoot Hodgkin (1910–1994). Born in Cairo, Dorothy spent her early years in Africa, particularly in the Sudan with her academic father, an educator and archaeologist, and her mother, also an archaeologist. Though she considered archaeology as a career, her love of chemistry developed at an early age, and she was one of only two girls permitted to study the discipline at her grammar school. She would go on to study chemistry at Oxford, and later go to Cambridge to complete her PhD. She came back to Oxford in 1934 to take on a research fellowship at Somerville College in chemical crystallography.

Crystallography is the experimental study of the arrangement of atoms in highly ordered microscopic structures, or crystalline solids. Hodgkin worked specifically on X-ray crystallography, which measures the angles and intensity of diffracted beams in a crystal or crystalline solid to determine the three-dimensional picture of it. Hodgkin is credited with developing the X-ray techniques needed to determine the structure of biologically important molecules and won a Nobel Prize for this work in 1964. These structures included cholesteryl iodine and the chemical formulae of penicillin and vitamin B12.

At the time Dorothy was working at Oxford, parts of the Department of Chemistry were still based in the museum. When the museum was built in 1860, it was home to all the science departments in the university, with each moving out over time as they grew, inevitably needing more space and specialized facilities. The Department of Chemistry remained in the museum for many years, probably because it was one of the few with a purpose-built laboratory – the Abbott's Kitchen – and had also had an extension in 1878. Hodgkin was based in the chemical crystallography lab in the museum, now used as a storage room for collections. The bust pictured here, sculpted by Anthony Stones in 2010 and recast in 2016, is located on a plinth along the east side of the museum court.

Nikolaas Tinbergen

To be awarded a Nobel Prize is one of the highest honours in the world and a number of Oxford scientists have achieved this accomplishment over the years. Once such prize was awarded to Nikolaas Tinbergen (1907–1988) in 1973 for Physiology or Medicine, along with Konrad Lorenz and Karl von Frisch, for their research on the organization and elicitation of individual and social behaviour patterns. Essentially this work looked at how animals respond to various stimuli using genetically programmed behavioural patterns, as well as the development of those responses and their origins from an evolutionary perspective. The research was critical to the development of a number of future discoveries, particularly in psychiatric medicine, including studies of obsessive compulsive disorders, stereotypic behaviours and catatonic posture. It was a breakthrough in understanding how behaviour functions physiologically.

Tinbergen often said he felt lucky to have spent his life doing what he loved so much – watching animals. Over a lifetime of work on animal behaviour he concluded that there were four key questions that needed to be asked in order to study this, and, more widely, the whole of the biological world. First was causation, or what stimuli led to a particular reaction and how learning affected this reaction. Second was development, which meant looking at how these behaviours change with age. Third was the function of the behaviour; does it have a benefit to the organisms' survival and reproduction? Finally he looked at behaviour from an evolutionary perspective by comparing it to behaviours in similar and closely related species and the possible mechanisms by which the adaptation was made.

Tinbergen's Nobel Prize medal, along with his other prize medals, was given to the museum in 1990 by the Tinbergen family. That year a temporary exhibition on his work was held in the museum and a conference commemorated his life and work. His research and discoveries continue to have an enormous impact on biological sciences to this day.

The Wollaston Collection

Standing at over six and a half feet tall, Thomas Vernon Wollaston (1822–1878) would have been an imposing sight, especially when wearing his top hat, as was the fashion in the mid-nineteenth century. Remarkable not only for his height but also for his entomological work, Wollaston was a prodigious taxonomist, describing hundreds of species of beetle from the Madeira and Canary Island chains as well as other Atlantic Ocean islands. He published numerous books and scientific papers, despite being plagued by ill-health for much of his adult life. His publications on the coleopterous fauna of Madeira and the Canary Isles are still considered the principal works for these regions to this day.

As part of his work Wollaston also assembled several large collections of insects, some of which are housed in the museum and are still consulted by scientists today. The collections reflect his meticulous approach and attention to detail, with each fragile specimen carefully pinned or glued to a card and colour-coded with its island of origin. These collections are a unique historical document of the fauna of the islands, some of which have been greatly altered by human activity and development

over the last fifty years. It is likely that many of the species in the collection are now threatened or extinct through habitat loss.

Wollaston was a contemporary and friend of many of the well-known names of the day in both entomology and, more widely, in natural sciences. He corresponded frequently with them, exchanging ideas and specimens. Many references to his work can be found in *The Origin of Species* but, although a friend of Darwin, he could not accept his theory of evolution by natural selection due to personal religious beliefs.

It is perhaps a great irony then that his collections and work reveal some of the fascinating evolutionary processes occurring on islands, such as adaptive radiation, where a species colonizing an island evolves rapidly to produce new species to fill vacant ecological niches. Wollaston himself communicated to Darwin the large number of flightless species of beetle he had observed on Madeira. This led to Darwin publishing his ideas on how flightlessness could evolve on islands.

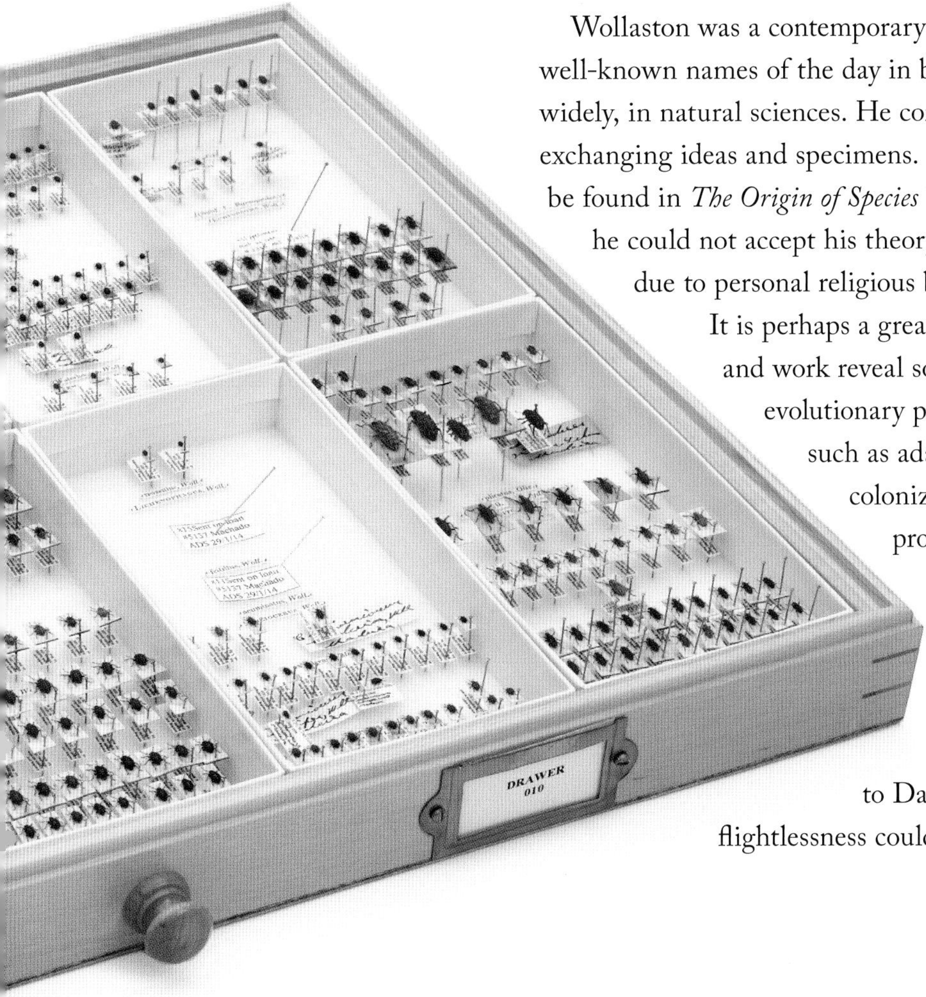

Mary Morland's minerals

Mary Morland (1797–1857), also known as Mary Buckland, was wife to Oxford professor, Revd William Buckland, Reader in Geology. She was a competent scientist, artist and collector of natural history specimens. Her work contributed to a number of very important and famous scientific publications, mostly credited to her husband. This work included her illustrations of the first described dinosaur remains, *Megalosaurus*. Many of these works are preserved in the museum's collections, but one such collection has proved to be a particular challenge.

The Mary Morland Mineral Collection comprises approximately 700 small hand specimens – of convenient size for study with the naked eye – of 'classic' British and European minerals, very typical of collections of that period. It was presented to the museum in 1997 by Drs John and Robert Eastwood, and had been acquired by their father, Dr Berwyn Eastwood, from a local antique dealer.

The provenance of the collection before this tells an interesting story. Around the time of World War II the collection had been acquired by a Lowestoft antique dealer and subsequently became the property of his nephew. Housed in a cabinet in the shop basement, only the topmost drawers of specimens escaped being submerged when Lowestoft was flooded by the sea in the storm surge of 1953. The rigours of that brief submarine existence left the collection in a poor state.

Its earlier history is recorded in two accompanying books which must have been in the top drawers. John Kidd's *Outline of Mineralogy*, 1809, has the bookplate of Sir Christopher Pegge, Regius Professor of Medicine at Oxford 1801–22, and a mineral collector. We believe the collector to be Mary as she spent much of

her childhood living in the Pegge household. Inside, Mary's eldest son Frank Buckland writes that the volumes were always kept in the cabinets of minerals in Buckland's drawing room in Christ Church and then at the Deanery in Westminster.

In addition, a catalogue of the collection, interleaved in a copy of William Phillips' 1837 *An Elementary Introduction to the Knowledge of Mineralogy*, was made by the London dealer James Tennant in 1871. That records the collection's origin as having belonged to 'Miss Morland, subsequently wife of Dean Buckland'.

As well as questions of provenance, the adventures of the Mary Morland Collection have resulted in other questions. The flooding resulted in the loss of most specimen numbers, and while many samples have been correlated with Tennant's catalogue, others are clearly missing, and to confuse matters further, a small number appear to be later additions, probably by Tennant.

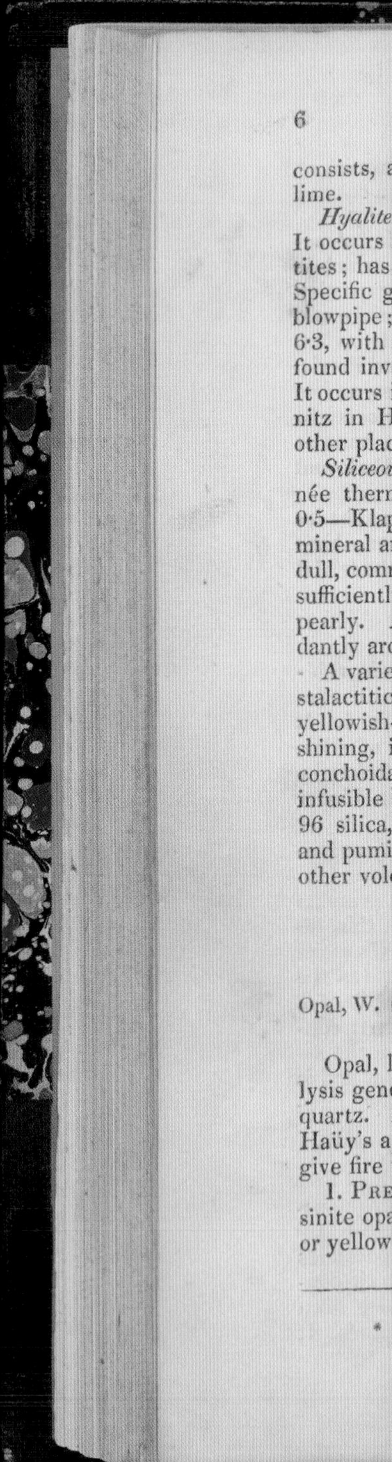

quelin, of 98 silica, and 2 carbonate of

ass. Quarz hyalin concretionné, H. nsparent botryoidal masses, or in stalac-e, is brittle, but is as hard as quartz. . It is infusible by itself before the ccording to Bucholz, of silica 92, water nina. This singular mineral is chiefly the cavities of trap or basaltic rocks. ear Frankfort-on-the-Maine, at Schem-nbedded in clinkstone at Waltsch and

elsinter, W. Quarz agathe concretion-nsists of silica 98·0, alumina 1·5, iron bout 1·8. The common colours of this n-white, and yellow. It is light, brittle, th a fibrous texture, although sometimes dmit of a conchoidal fracture; lustre before the blowpipe. It occurs abun-osited by, the hot springs of Iceland. *Pearl sinter* or *Fiorite*, which occurs in otryoidal, and globular masses, of a white, sh colour; externally it is smooth and ning with a pearly lustre; fracture flat n the edges; not so hard as quartz; and vpipe without addition. It consists of d 2 lime. It occurs in volcanic tufa tine; the Florentine dominions; and in Italy.

OPAL.

I. Silex Opale, Bt. Uncleavable Quartz, M. Sp. Gr. 2·09.

ists chiefly of silica and water; but ana-a greater quantity of the latter than in esinous lustre is probably the origin of e of its varieties are sufficiently hard to

OBLE OPAL. Edler opal, W. Quarz re-beautiful mineral is of a white, bluish, , and when viewed by transmitted light

in allusion to its glassy appearance.

69. Amethyst, a detached hexagonal pyramid
Minas Geraes, Brazil.

70. Do with polished planes,
same locality

71 Do hexagonal pyramid,
Co. Cork, Ireland.

72. Do hexagonal prism, with the pyramid in alternate large and small planes, Minas Geraes, Brazil.

73. Do on massive quartz,
near Redruth, Cornwall.

74. Do with rounded pyramids,
Porcura, Transylvania.

75. Do as a vein between bands of Agate
Oberstein

76. Do in successive superimposed bands of crystals each coated with white quartz
Schemnitz, Hungary.

77. Do polished stones

78. Do facetted stones

79. Do a fragment exhibiting the rippled fracture considered to be peculiar to this variety, Minas Geraes, Brazil.

80. Yellow Quartz or Citrine of the jewellers
Minas Geraes, Brazil.

81. Do a darker variety,
same locality.

82. Do in fragments somewhat milky, and exhibiting the rippled fracture of amethyst same locality.

83. Do facetted stone.

Hands on the otter

The museum has been a sector leader in the use of touchable specimens within its public galleries. Since the early 2000s it has been policy to look for opportunities to incorporate touch in all new visitor experiences, from events to exhibitions, and to support research into this form of collections use.

Taking away the glass case and removing the barrier has provided a new avenue of exploration in the museum. Being able to touch and engage with a specimen in this memorable way – stroking its fur, feeling the shape of its face or the hardness of its teeth – is particularly important for blind and partially sighted visitors. Specimens from across the collections have been used to illustrate the natural world in this way, from sparkling minerals to smooth snake skin.

All taxidermy is specifically commissioned so that it is suitable for handling by visitors. This can be challenging as much material is fragile or, as in the case of the thorny shark, difficult to handle safely. Objects need to be robust enough to withstand the touch of many fingers, and even the occasional grab from smaller hands. Each one needs to be assessed and undergo specialist conservation work to ensure it looks its best and can handle the pressures of celebrity status.

All material is, in addition ethically sourced or donated, such as the otter that now inhabits one of the purpose-built tables in the Sensing Evolution display. This otter was a victim of a road traffic accident and is a sad but important reminder of one of the ongoing conflicts between people and wildlife. The terrapin that stands nearby has a happier story. A well-loved pet, Terri was donated to the museum by her former owner after her death so that she could be part of the touchable display and continue to teach and enthral people for many years to come.

Hard as snails

In deep-sea hydrothermal vents in the Indian Ocean at depths of more than 2,500 metres, just beside the black smokers that are churning out superheated water exceeding 350°c, lives *Chrysomallon squamiferum*, a species of scaly-foot snail. It is the only known gastropod with a suit of scale armour, the scales and shell being mineralized with iron sulphide.

Scaly-foot snails were first discovered in 2001, at the Kairei vent field in the Indian Ocean. The discovery came as a great surprise: the adaptation was remarkable, even when compared to other animals specialized for living at vents. Although the shell of a snail is well known to be modified into a great variety of forms, no other gastropods have mineralized structures on the foot, yet this species has thousands of scales. There are several theories about the function of the scales – for example, they could be for either protection or detoxification – but their true use remains a mystery.

Hydrothermal vents were first discovered in the Galápagos Rift as recently as 1977. This is just off the Galápagos Islands, whose fauna famously inspired Charles Darwin in the development of his theory of natural selection. Vents are deep-sea 'hot springs' fuelled by geological activity; the hot erupting fluid is usually acidic and contains various metals, as well as hydrogen sulphide. This is what makes rotten eggs smell bad, and is toxic to most organisms. Some bacteria, however, are able to use it to produce energy in a process known as chemosynthesis.

Over geological timescales many remarkable organisms have adapted to live in these toxic utopias, and flourish by exploiting the energy produced by these bacteria. The scaly-foot snail has also harnessed the power of chemosynthesis, housing endosymbiotic bacteria – bacteria living inside another creature to mutual benefit – in an enlarged part of its gut. This internal food factory produces the energy it needs and this notable adaptation is likely to be the reason it is three times larger than its closest relatives.

Despite being known to science for over a decade, the species was not formally described until 2015. The museum has since received a set of five specimens and other specimens have been deposited in a number of institutions around the globe as part of the description process, which will serve as key references for scientists who wish to study this extraordinary species in the future.

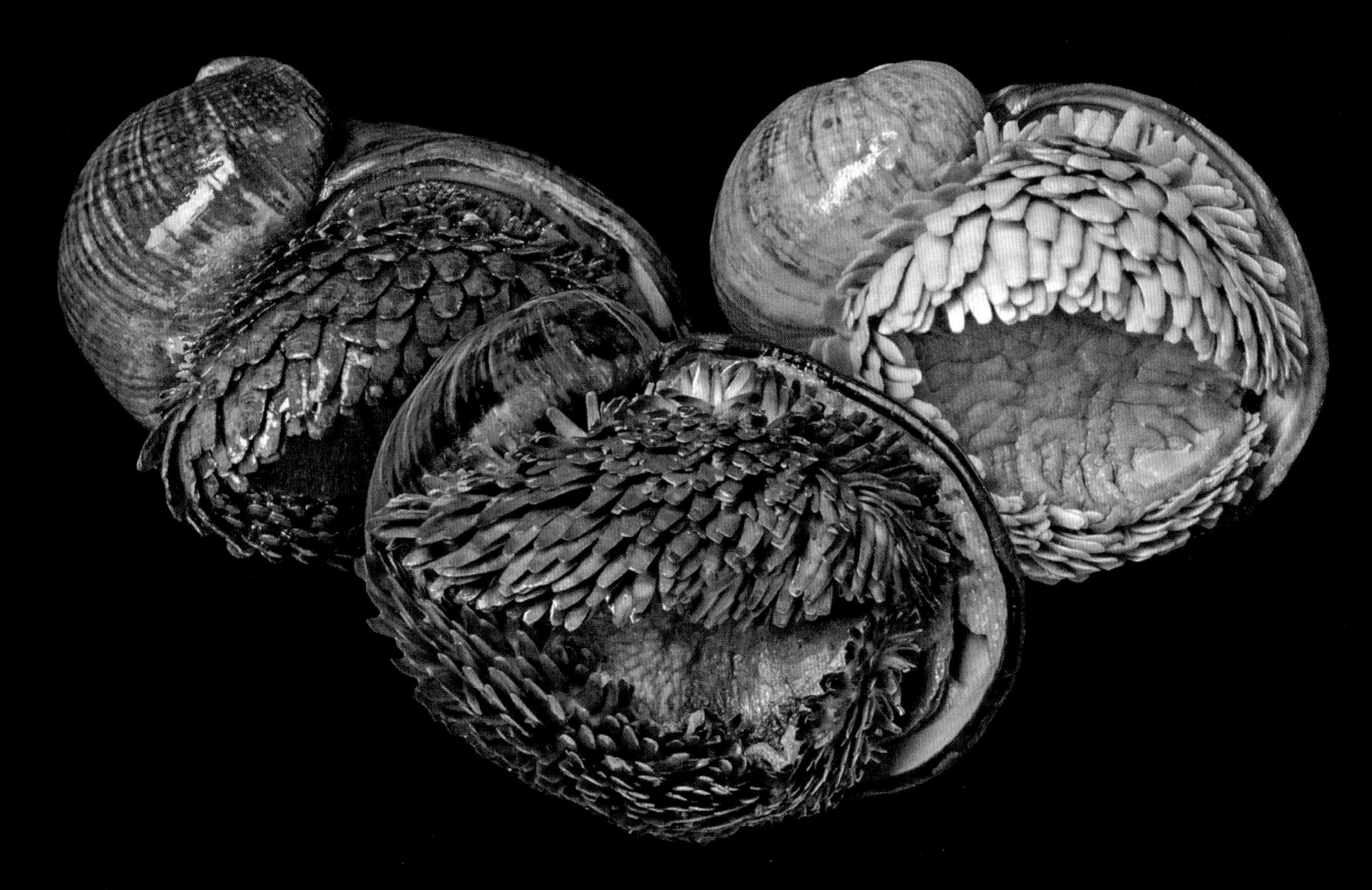

Skull of *Stratesaurus*

Plesiosaurs were one of the most successful groups of aquatic tetrapod, or four-limbed vertebrates, to have existed. They lived over a geological history spanning 135 million years and were distributed globally, with over a hundred species known from across the world, including Antarctica. Adults could grow up to fifteen metres, with the head size and neck length varying greatly across the group. For example, one species had a neck longer than its body and tail combined, with seventy-six neck vertebrae – more than any other animal that has ever lived. A diverse array of dietary preferences has also been inferred, including large reptiles, fish and cephalopods, bivalves and crustaceans. Although plesiosaurs are mostly found in marine deposits they are also represented in freshwater lake, lagoonal and near-shore deposits.

This is a beautifully preserved skull of a small plesiosaur called *Stratesaurus taylori*. It is the holotype specimen and was named by R.B.J. Benson, M. Evans and P.S. Druckenmiller in 2012. It was discovered in Street, a village in Somerset, and donated to the museum, along with a large collection of other Jurassic marine reptiles, by Thomas Hawkins in the late nineteenth century.

This specimen is one of the world's smallest plesiosaurs. In total it would have been about two metres long from the tip of the snout to the end of the tail. It is also one of the world's oldest plesiosaurs, from the earliest Jurassic, 200 million years ago. For

many years, small plesiosaurs from this part of Somerset were all referred to as the same species, *Thalassiodracon hawkinsii*, but recently researchers realized that there were actually four species represented. Because these plesiosaurs have similar overall body plans and small body size, it is difficult to distinguish between them. A beautifully prepared specimen like this, however, shows the anatomic details very clearly, allowing the new species to be recognized for the first time.

Whale of a tale

Five skeletons of whales and dolphins have hung from the roof arches of the central exhibition space since the early years of the museum. Acquired for their scientific research potential and for use in anatomy courses, the specimens were presented to the museum by some of the most important cetacean researchers of the late nineteenth century, such as Danish zoologist Daniel Frederik Eschricht, who gave the museum two of the specimens now on display: the humpback whale skull at the entrance and the suspended minke whale skeleton.

The bottlenose dolphin was caught near Holyhead in 1868 and was drawn by another notable natural historian, William Henry Flower, before being skeletonized for the museum. The orca skeleton is from an animal killed in the Bristol Channel by fishermen in 1872, and the beluga whale was collected from Spitsbergen, Norway, in 1881, and presented by Alfred Henge Cocks.

When the roof renovation project was undertaken in 2013 it afforded a unique opportunity for the museum's conservators to access the specimens and undertake some much-needed conservation work. The specimens were lowered and the scaffolding for the roof renovations constructed around them in such a way as to form a safe pocket of space, visible from ground level, from where members of the public could watch the conservation process.

Years of light exposure and the accumulation of dust on the bones had led to them becoming cracked and dirty. The internationally recognized 'Once in a Whale' project undertook to change this and the transformation has been spectacular.

Through an extensive treatment programme by a team of conservators, the years of built-up dust and oil seepage were painstakingly removed. The bones were then consolidated, a process by which the porous bones are injected with resin to bond friable pieces and keep brittle sections intact.

Once the specimens had been cleaned and treated they were rearticulated, ensuring that they are both anatomically correct and preserved for future generations of visitors to enjoy.

Mbarikuku or moustached kingfisher

The moustached kingfisher, or Mbarikuku to give it its native name, is a species of endemic kingfisher from the Solomon Islands. This bird is poorly known to western science. There are some scattered records of sightings and observations made by the few field scientists who have visited the region over the centuries, but otherwise little is documented other than the fact that the bird inhabits the more remote and difficult-to-reach areas of the mountain ranges. The first specimen to be collected was in 1927 when a single female was taken and later described as the holotype for the species. Two further female birds were prepared by members of an expeditionary team from Oxford University after they had been collected by local hunters. The lack of material in museums or published information at that time led to a perception that the species was rare. Modern research undertaken in 2015 has established that a small but healthy population exists in the islands today.

During this recent research a single male specimen was collected to compliment the material already held in museums. From these four specimens it should now be possible to collect a comprehensive data-set for morphological and molecular work. Old preservation methods have meant that this work was previously either limited or impossible due to the condition of the specimens. Chemicals used in the preparation of skins were often toxic and could degrade genetic material. The male of the species can also now be formally described in the literature in order to give a more complete species description. Most importantly, scientists will now be able to study change through time, comparing historic material with recent specimens.

When the male was collected in 2015, it caused quite a controversy, since many people were seemingly shocked that museums still collected biological specimens. Birds, mammals and other charismatic megafauna tend to elicit stronger emotional responses than marine invertebrates or insects. Chris Filardi, a member of the joint research team from the Solomon Islands and American Museum of Natural History, wrote a heartfelt article defending collection; making museum specimens available for long-term verification and follow-up work as analytical techniques improve is a core function of museum collections in understanding the natural world.

A plesiosaur named Eve

Major discoveries and important finds are often exclusively seen as the domain of scientists and experts. Within natural history, however, individuals and groups from outside academic institutions have long played a key role in the development of knowledge of our natural world, and these kinds of contributions continue to this day.

One such example is the recent discovery of a plesiosaur, a Jurassic marine reptile, nicknamed Eve. Eve was discovered in 2014 by a group of dedicated avocational palaeontologists called the Oxford Clay Working Group in a Peterborough quarry. Though Oxford Clay, which is a type of rock found across the UK, is a rich source of beautifully preserved fossils, finds like this are rare today

as most excavation of clay is now completely mechanized, meaning fossils are often destroyed before they can be discovered. The group decided to call their find Eve, as it was their first major discovery. We still do not know for certain whether it is male or female.

Eve is an incredibly rare find: a nearly complete fossilized skeleton, including the skull, vertebrae and flipper bones. After the discovery in November 2014, the group began excavation. Over the course of four days, the group dug up over 600 separate pieces of bone. Carl Harrington, the member of the group who first spotted the specimen, then spent over 400 hours cleaning the specimen and gluing most of the pieces back together.

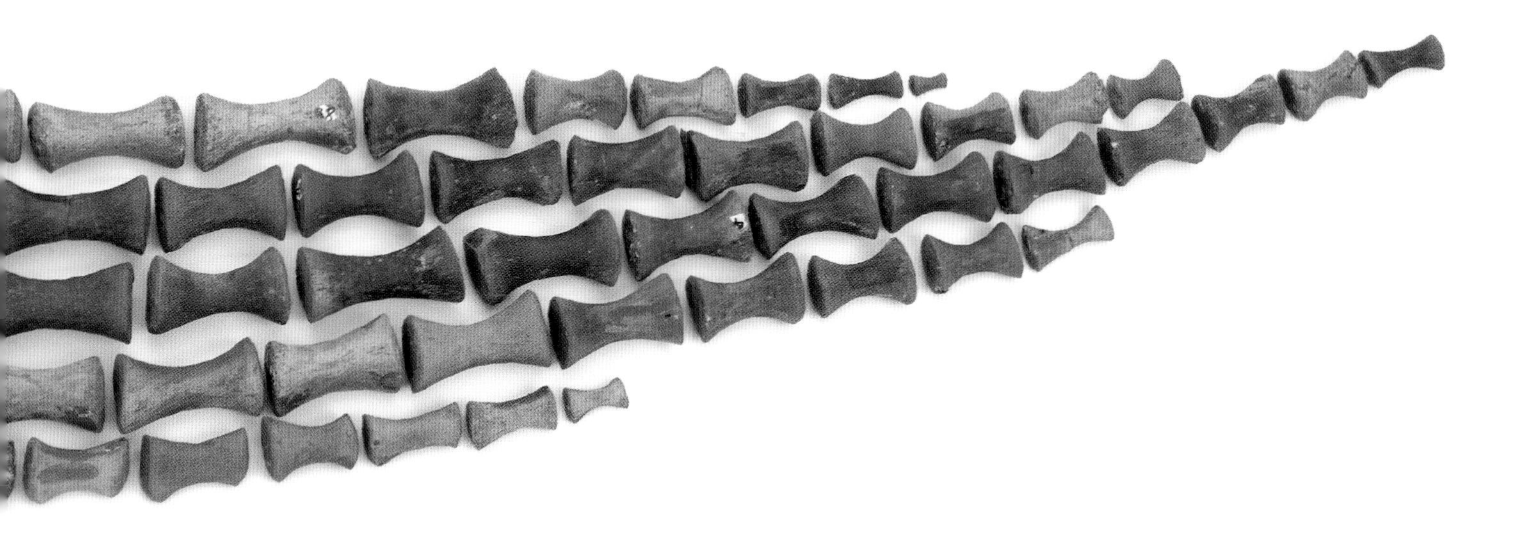

In 2015, Eve was donated to the museum by quarry owners Forterra. Since it has been with the museum it has been further prepared and studied, with preliminary research suggesting that Eve is a new species of plesiosaur.

The skull is still inside a block of clay (shown here) and is being carefully removed by the museum's preparator. A CT scan of the clay block was made and the skull reconstructed in 3D while still inside. This allows the preparator to extract the bones more easily as she knows where to find them. She compares it to having a picture of the puzzle on the lid of the box. Some parts of Eve, including most of her backbone, are preserved in extremely dense nodules that cannot be prepared without damaging the bone. New micro CT scanning technology at the University of Warwick will enable staff to see the bones inside these nodules and reconstruct them virtually.

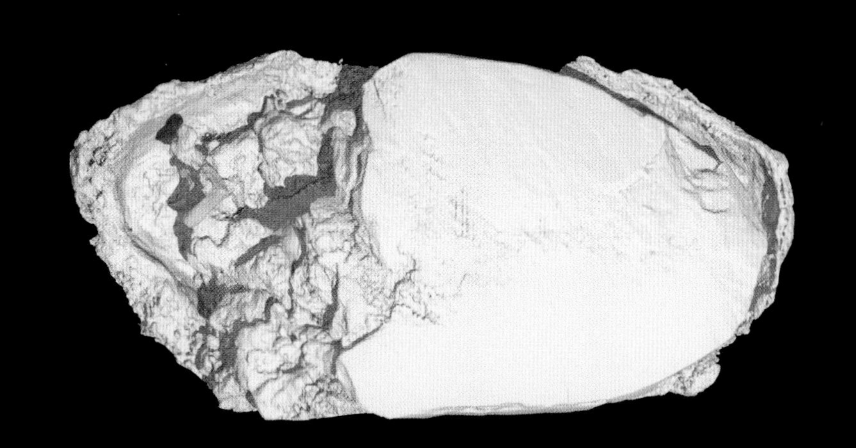

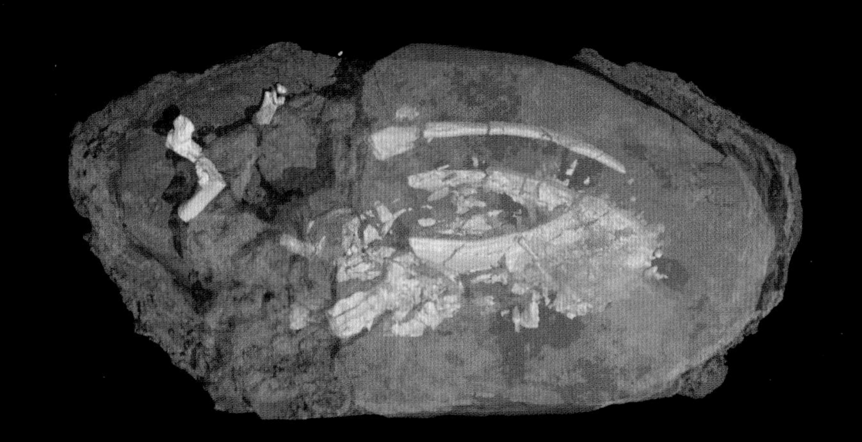

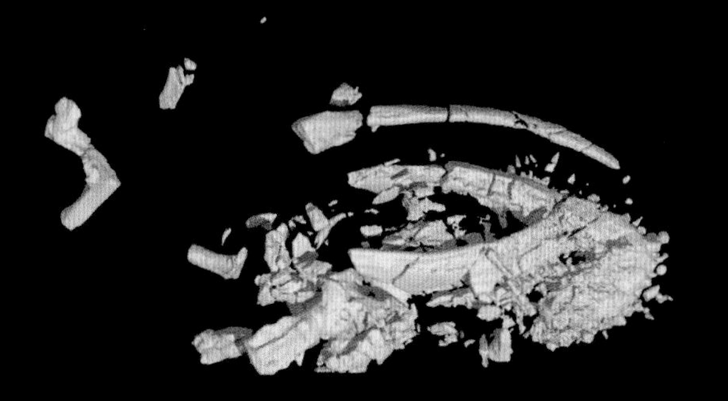

Pistol shrimp

Taken as whole, the Crustacea Collection is one of the jewels in the museum's crown. Built upon historic donations of material from researchers such as Charles Darwin, and in more modern times through years of research by Dr Sammy De Grave, one of the world's foremost experts on decapods, it contains much in the way of type specimens and examples of rare species. The collection is primarily a taxonomic masterpiece used for morphological and genetic research, though held inside the numerous jars and tubes filled with spirit preservative lies a vast wealth of information relating to biodiversity, species distributions and habitat loss, environmental contamination and global climate change.

The order Decapoda encompasses crabs, lobsters, crayfish, prawns and shrimp, the majority of which are marine animals. Perhaps best known for their economic importance in commercial fisheries they also have essential ecological roles. Found in habitats around the world, from mangrove swamps to deep sea hydrothermal vents, decapods demonstrate some amazing diversity and adaptation. The recently described pistol shrimp *Synalpheus pinkfloydi* is stunning, both literally and figuratively. It uses its beautiful bright pink claw to ward off predators and stun prey by snapping it shut at high speed, producing a shock wave that is one of the loudest sounds in the ocean. *Periclimenes rincewindi* was named for its ability to blend in with its surroundings. This little shrimp rides around on a species of hair star, a kind of free-swimming starfish. It has evolved a unique colour pattern that matches its host so well that it was only identified and described in 2014.

The museum collection currently holds approximately 1,177 species of shrimp. This databank of information can be used for a wealth of research purposes and the museum shares both information and specimens freely with researchers all over the world. A recent large-scale study of freshwater shrimp found that almost a third of all species are 'Threatened' or 'Near-Threatened' according to the International Union for Conservation of Nature (IUCN) Red List criteria. A team of fourteen authors, including Dr De Grave, worked collaboratively to produce the first global assessment of extinction risk for this group, taking data from the collections and pooling expertise. The importance of this work and of maintaining such collections places the museum on a world stage, and is just one of the many roles that it plays in supporting our understanding of the natural world.

The evolution of echinoderms

One of the most iconic seashore animals is the starfish, probably because of its unusual shapes and often bright colours. Together with other well-known sea animals such as sea urchins and sea cucumbers, starfish are characterized by a unique form of symmetry called fivefold symmetry, or pentamerism, meaning that they are divided into five roughly equal parts. While fivefold symmetry is abundant in the natural world, particularly in plants, what makes these animals unusual is the fact that they begin life with bilateral symmetry, or in two equal halves, like most animals. Named echinoderms, they are one of the only groups of animals to feature this unusual trait. The closest living relative to echinoderms, worm-like creatures called hemichordates, also begin life with bilateral symmetry.

From an evolutionary perspective this is a very interesting observation. Echinoderms have an excellent fossil record because they possess a hard, mineralized skeleton, which greatly enhances their chances of being preserved as fossils compared to soft-bodied organisms. These fossils document the earliest history of echinoderms, and so help us to better understand their evolution. The first fossil echinoderms are over half a billion years old, and include extinct groups that show both bilateral and fivefold symmetry. In addition, some fossils exhibit threefold symmetry, and some lack a clear plane of symmetry – in other words they are asymmetrical.

Based on our understanding of living animals, and using modern methods for reconstructing the relationships between different species, it is possible to infer that the early fossil echinoderms with bilateral symmetry belong at the base of the echinoderm evolutionary tree. The next branches in the tree lead to the asymmetrical fossil groups, and these are followed by those forms that show threefold symmetry. Lastly, we see the diversification of forms with fivefold symmetry, including species belonging to the groups that still exist today, such as the starfish. The fossilized crinoid or sea lily illustrated here shows two of its five branched arms in side view.

Using the fossil record, we can see a clear picture of how echinoderms evolved from worm-like organisms into star-shaped creatures.

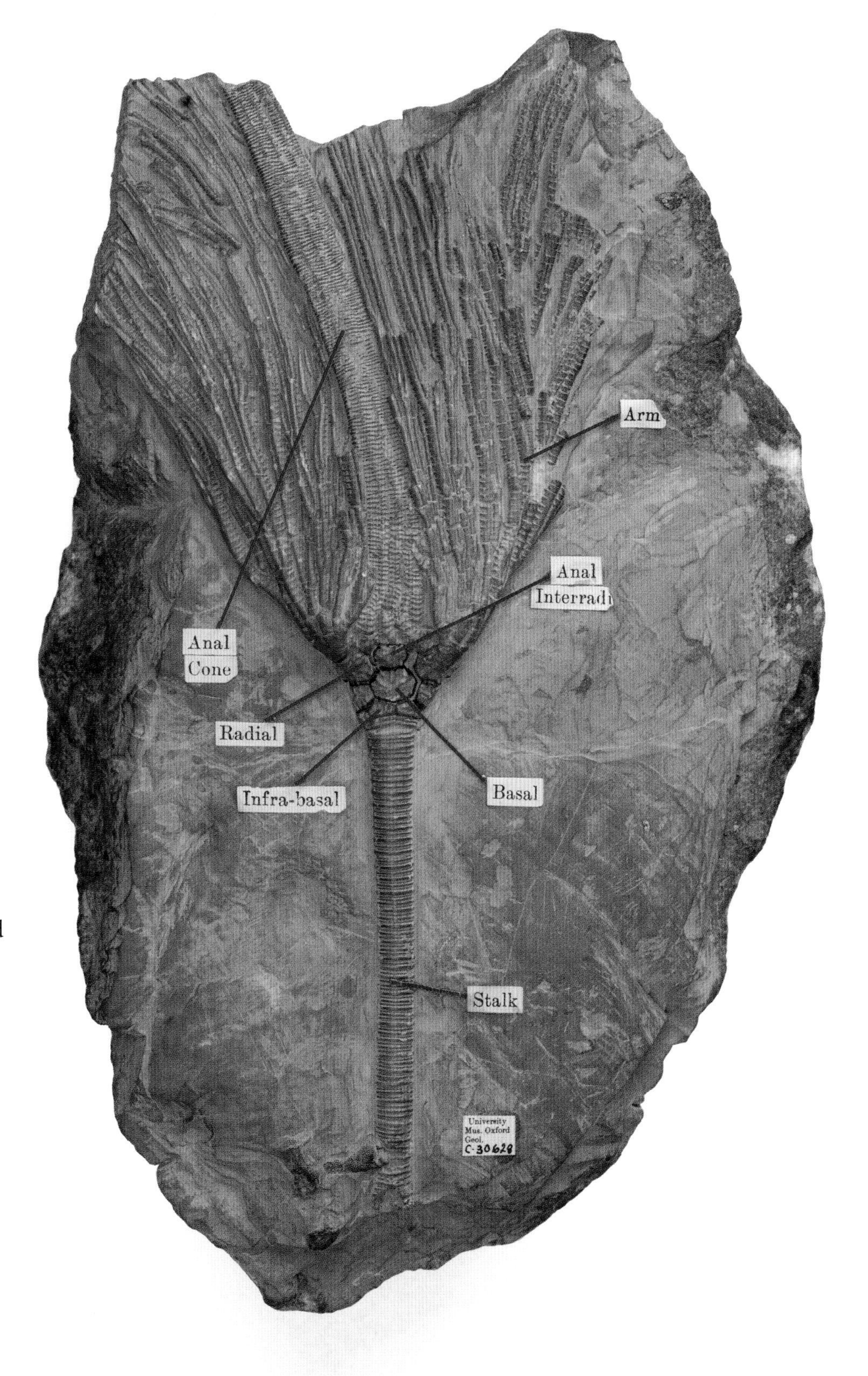

Crabs in the coral

In the eighteenth and nineteenth centuries several American and European ship-based expeditions collected stony corals for the museums in their home countries. Corals were rather easy to collect, especially those species living in shallow water, and once dried and cleaned the skeletons could easily be stored and studied. Many of the currently known coral species were scientifically described based on these historical collections, mostly by scientists who had never seen a living coral reef with their own eyes.

Corals are also home to a wide range of small creatures that inhabit nooks and crevices in the coral skeleton. The holes and cavities in corals might have been considered random damage by the early scientists, but are now thought to have been made by symbiotic organisms. One group of tiny crabs is completely adapted to living in dwellings in corals. These 'gall crabs' are at the heart of the work of Museum Research Fellow Dr Sancia van der Meij. She studies how animals live together on coral reefs, and uses the gall crabs as a model group.

Natural history collections are an incredibly rich source of information for this type of research. The dwellings that these gall crabs inhabit are unique in shape and size, hence their presence can be detected in historical collections. This has led to several collection-based scientific discoveries. Based on fossil coral collections, the presence of gall crabs in the Pliocene and Pleistocene eras in the Western Atlantic can be reconstructed. Such details help scientists to understand how coral reef organisms evolved and the complex nature of the ecosystem they inhabit.

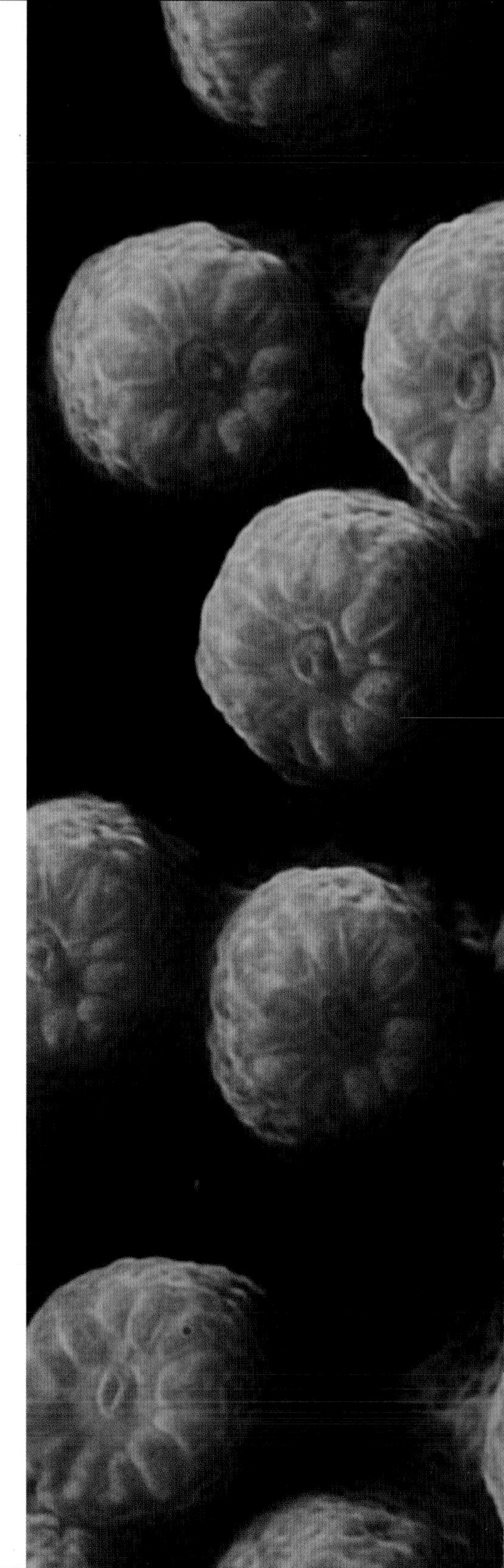

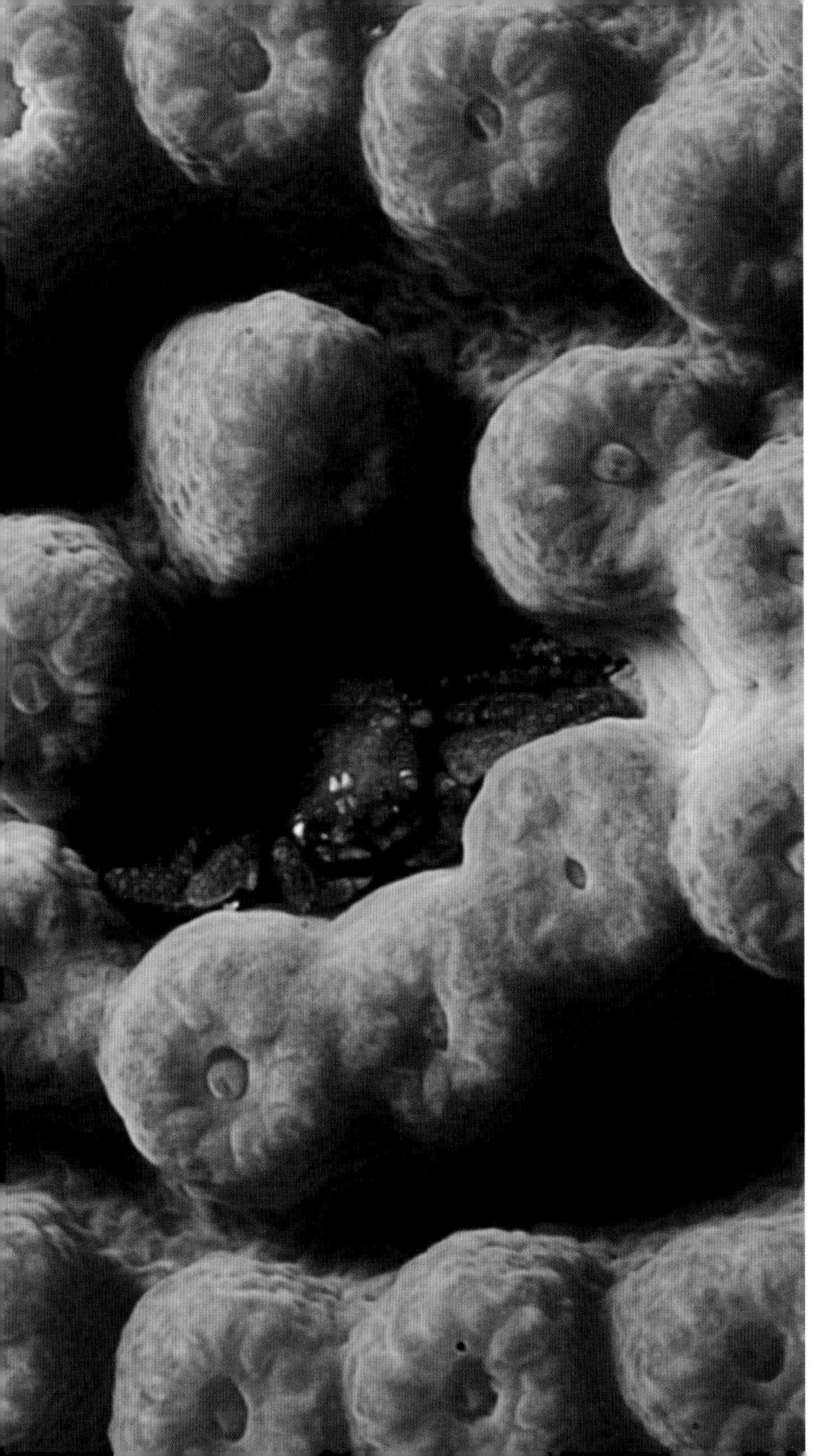

The dry coral collections in Oxford University Museum of Natural History mainly originate from the Sudanese Red Sea and adjacent regions. This area is not very accessible for scientific research, but based on the museum's coral collection the first records of gall crabs from Sudan have been identified. Current scientific research allows museums to reinterpret their historical collections, giving new meaning and value to their objects and collections.

A shattered skull

Sometimes the specimens in the collections can tell a very personal story. Recent work by Dr Kathryn Krakowka, a human osteology researcher who has been analysing bones from the museum's collection of human remains, has uncovered one such tale.

The museum holds many specimens that are not on public display, and perhaps one of the largest collections hidden from view is that of human remains. Like many of the oldest specimens at the museum, the collection first started at Christ Church. In the eighteenth century, Dr Matthew Lee began collecting specimens to be used for medical teaching, and upon his death he bequeathed parts of the collection and a financial endowment for a Readership in Anatomy to his alma mater, Christ Church. Each subsequent 'Lee's Reader in Anatomy' continued to add to the collection of medical specimens, with the largest contributor being Henry Acland (1815–1900). As one of the founders, Acland brought

his anatomical and physiological collection with him when the University Museum was opened in 1860.

The collection continued to be added to throughout the nineteenth century, with members of the university often bringing back specimens from their expeditions. By the end of the century it had grown to contain over 1,400 specimens.

The recent project has uncovered some interesting specimens, such as the calotte or top portion of the cranium of an eighteenth-century Oxford man. The calotte has large parts of the frontal and parietal bones missing. The margins around the missing portions are smooth, suggesting that this was a traumatic injury that had fully healed. The inside of the cranium provides some explanation for this, as attached to the surface of the bone are three labels that detail the story of this individual:

> The individual to whom this skull belonged was buried for seven hours under a heap of stones which fell in upon him from the side of a well that he was clearing out in St Mary Magdalene parish in Oxford. Sir Charles Nourse saw him immediately after his removal, he continued to attend him during three months; in the course of which time the large piece of bone … came loose. From the pain [occasioned?] by the loose bone the man required Sir C. Nourse to take it out; but Sir C declined this saying it would kill him. The man however himself sitting down before a … cut it out with a razor. There afterwards the other … becoming loose Sir C took … of the man [completely recovered?] by a year and a half. To pursue this unusual occupations he was in the habit of attending the Anatomical Lecture for the purpose of exhibiting himself during four years; when he died. The accident happened about the year 1760; Dr Smith, Vice Principal of the St Mary Hall (according to Thomas Knapp's account) obtained the skull and bones.

The Charles Lyell Fossil Collection

One of the least known but most impressive fossil collections held by the museum is that belonging to Charles Lyell (1797–1875). The collection contains over 16,000 specimens comprising primarily Cenozoic (66 million years ago to present) molluscs from Europe and North America. It includes bivalve species similar to clams, oysters and mussels, as well as gastropods, or snail-like creatures. The collection was given to the museum by Lyell's nephew, Sir Leonard Lyell, in 1903. Its importance has, however, less to do with the specimens and more to do with their famous collector.

Charles Lyell is best known for his book *Principles of Geology*, which popularized and extended James Hutton's concept of uniformitarianism – the idea that the Earth was shaped by the continuous and uniform processes still operating upon it today (see also p. 85). He was a close and influential friend of Charles Darwin, and Lyell's *Principles* was one of the few geological books that Darwin took with him on his voyage on HMS *Beagle*. Volume 1 of the first edition, published in 1830, was given to Darwin by the ship's captain shortly before they set out, and volume 2, published in 1832, was sent out to him in South America.

For a long time Lyell failed to share Darwin's belief in evolution by natural selection. Only during the final revision of *Principles of Geology* in 1865 did he fully adopt Darwin's conclusions. Why Lyell was hesitant in accepting evolution is best explained by Darwin himself: 'Considering his age, his former views, and position in society, I think his action has been heroic.'

In addition to his influence on Darwin, Lyell's other major scientific contribution was the division of the Tertiary period of geological time into the Pliocene, Miocene and Eocene epochs, based on the increasing similarity of the fossils to present-day fauna. His fossil collection in Oxford contains the material on which this work was based.

The museum has recently completed a project which has seen this important collection catalogued, and it is now available online.

Stacked insects

Insects are notoriously difficult to photograph and this is in large part due to their small size. Of the million or so described species the majority are under five millimetres. They also have a wide variety of morphology, with horns, spikes, nodules, hairs, pits and more, creating intricate three-dimensional structures that add layers of complexity when attempting to image specimens. The need for good-quality pictures is ever increasing, however, as photographs are becoming an integral part of both collections and research work. They act as valuable records of specimen condition for museum staff, who then provide those images to researchers to use for morphological work. This latter use is of particular importance for historic type specimens, many of which are very fragile and can no longer be posted safely. High-quality images can act as a useful surrogate for those researchers unable to travel to Oxford to see the specimens in person.

Images such as the ones seen opposite and overleaf are created through a camera attached to a microscope. Multiple pictures of a specimen are taken on a variety of different focal planes. The resulting set is then processed through a specialist piece of software that analyses each picture and creates a compound image, stitching together the sections of each so that the resulting photograph shows the insect completely in focus from the tips of its antennae to the end of each claw.

These images are not only a valuable resource for staff and researchers but are beautiful, uncovering the fine details, stunning colouration and micro-sculpture of each species. Visually arresting and aesthetically pleasing, they have become something more than a mere catalogue record, being a piece of art in their own right.

The end of a bird

The dodo is the most iconic specimen in the museum: it even forms part of the museum's logo. But as well as the biological remains, the museum also holds one of the oldest and best-known depictions of the famous extinct species.

Jan Savery the Younger (1589–1654) was a Dutch Golden Age painter. He was the son of Jacob Savery and nephew to Hans Savery the Elder and Roelant Savery: all three were painters. He was likely a pupil of his uncle Roelant, who painted as many as ten depictions of the now extinct bird during his career.

Jan Savery is best known for his 1651 depiction of the dodo now held by the museum. The painting was presented to the Ashmolean in 1813, before the University Museum had been built, by W.H. Darby. The animal illustrated appears to have downy feathers and a relatively large head, features usually associated with younger birds, but, as Jan's uncle had been known to depict the dodo with webbed feet, there was obvious confusion about the bird's appearance. A few dodos were brought alive to Europe in the early 1600s. Confined without exercise and given the wrong food, they became obese. This may explain the difference between its appearance in the painting and what we now believe the dodo looked like in the wild – a much more slender bird.

It is widely believed that Lewis Carroll, an Oxford mathematician and author of *Alice's Adventures in Wonderland*, was inspired by Savery's image of the dodo hanging in the museum to include the funny-looking bird as a character in his book. Many artists, scientists and writers continue to be inspired by the museum's collections to this day.

Above our heads

The most notable architectural feature of Oxford University Museum of Natural History is its stunning glass roof. When the competition was held to build the university's new museum, which was to be a 'cathedral to science', the roof was one of the key elements that persuaded Acland and Ruskin to select Benjamin Woodward's design (see p. 14). The roof was also a feat of engineering genius – though it has caused some issues in its 157-year history.

During the construction of the University Museum, the original design primarily used wrought iron to hold the weight of the glass roof and its wooden support structure. The pillars were to be elaborately decorated, in keeping with the Pre-Raphaelite-inspired design of the interior fixtures and fittings. Unfortunately, the calculations proved to be faulty and the roof collapsed shortly after work began. Wrought iron was too weak a material to hold the great weight of the roof and E.A. Skidmore, an experienced ironmaster, was brought in to produce a cast-iron design to be decorated with wrought iron.

Another challenge presented by the roof is the conditions it produces within the museum exhibition space. Our understanding of the effects that certain environmental conditions have on museum collections has changed considerably over the past 150 years. We now know that prolonged light exposure and fluctuating temperatures can harm artworks and archival material, as well as taxidermy and spirit collections. With advances in technology and improvements to cases the building and collections can be preserved for future generations to enjoy. The roof is still one of the museum's greatest treasures and a perfect example of a complex problem leading to innovation and improvement.

IRON SKIDMORÉ COVENTR

Object numbers

Individual specimen/object numbers are preceded by OUMNH (Oxford University Museum of Natural History)

PAGES	OBJECT	ACCESSION/OBJECT NO./SHELFMARK
16	[*Entrance Porch Archway Design*] by John Hungerford Pollen and Thomas Woolner [attributed], date unknown	History of the Building of the Museum Collection
17	[*Entrance Porch Archway Design*] by Thomas Woolner, date unknown	History of the Building of the Museum Collection
18–19	Oxford dodo, *Raphus cucullatus*	ZC.11605
20–21	Ammonite, *Asteroceras obtusum*	J.1194
22	*Archetypa studiaque patris Georgii Hoefnagelii* by Jacob and Joris Hoefnagel, 1592	S.3
24–7	*The History of Four-Footed Beasts and Serpents* by Edward Topsell, 1658	50 c. 20
28	Common warthog, *Phacochoerus africanus*	ZC.17499
31	*A Natural History of Oxfordshire* by Robert Plot, 1677	AL YZ/105
33	Bath white butterfly, *Pontia daplidice*	J.C. & C.W. Dale Collection
35–7	*Metamorphosis Insectorum Surinamensium* by Maria Sibylla Merian, 1705	E.32
38–41	*The Aurelian or Natural History of English Insects* by Moses Harris, 1766	E.21
42	Krasnojarsk or 'Pallas Iron' meteorite	Mt.001
43	Nakhla meteorite	Mt.084
44–6	*Jones' Icones* by William Jones of Chelsea, date unknown	WJ/B/1
48–9	Letter to William Jones from James Smith, 7 July 1787	WJ/A/1/1/003
50	Ham beetle, *Necrobia ruficollis*	J.J. Walker Collection
51	*Pierre André Latreille*, by Louis-Parfait Merlieux, 1833	Museum Art Collection
52–3	Giant Irish elk, *Megaloceros giganteus*	Q.1972
54–5	*A Section of the Strata of Derbyshire Forming the Surface from Alton Eastward of Ashover Southwardly through Bonsal beyond Grange-Mill* by White Watson, 1822	PAL.T.00002
55	*Sections of the Limestone Rocks Described in Watson's 'Delineation of the Strata of Derbyshire'* by White Watson, 1821	PAL.T.00001
56–7	Early mammal, *Phascolatherium bucklandi*	J.20077
58–9	*Megalosaurus bucklandii*	J.13505
60–61	*Under Jaw and Teeth of Megalosaurus* by Mary Morland, 1824	William Buckland Print Collection, WB/A/D/132
62–3	*The Geological Lecture Room, Oxford* by Nathaniel Whitlock, 1823	Museum Print Collection
64–5	[*Landslip at Lyme*] by Mary Buckland (née Morland), c.1840	William Buckland Print Collection, WB/A/Landslip/1
66–7	Hippo tusk	ZC.19175
	Ground beetle	W.J. Burchell Collection
	Lily weevil	W.J. Burchell Collection
	Swallowtail butterfly	W.J. Burchell Collection

	Antlion	W.J. Burchell Collection
68-9	*Inside Mr Burchell's Waggon* by William Burchell, 1820	William Burchell Collection WJB/D/1/1/001
70-71	*A Delineation of Strata of England and Wales* by William Smith, 1815	WS/H/1/0/001
72	*Crocuta crocuta spelaea*, left maxilla	Q.6117
72-3	Illustrations from *Reliquiae diluvianae: or, Observations on the organic remains contained in caves, fissures, and diluvial gravel, and on other geological phenomena, attesting the action of an universal deluge* by William Buckland, 1823	
73	*Crocuta crocuta spelaean*, lower left carnassial	Q.6106
74-5	Red Lady of Paviland, *Homo sapiens*	Q.1
76-7	Right femur of *Cetiosaurus oxoniensis*	J.13615
78-9	Decorative rocks:	
	Granito bianco e nero, from Wadi Barud, near Mons Mlaudianus, Eastern Desert, Egypt	no. 844
	Rosso antico, from Akra Tainaron (Cape Matapan), Mani peninsula, Laconia, Peloponnese, Greece	no. 62
	Verde antico, from Larissa, Thessaly, Greece	no. 565
	Sicilian jasper, diaspro di Palermo, probably diaspro di Navurra, from Palermo, Sicily, Italy; most probably from Feudo Navurra, near Casteldaccia	no. 776
	Most probably giallo e nero di Carrara, from Carrara, Massa e Carrara, Tuscany, Italy	no. 131
	Breccia di Settebasi, from Isle of Skyros, Sporades, Central Greece	no. 405
	Alabastro di Palombara, possibly from Iano di Montaione, Siena, Tuscany, Italy, but more probably Pamukkale, Denizli, Turkey	no. 311
	Cipollino verde, from near Karystos, Euboea, Central Greece	no. 90
	Lapis lazuli, from Kokcha Valley, Badakhshan, Afghanistan	no. 703
	Porfido serpentino antico, from Krokees (Levetsova), Laconia, Peloponnese, Greece	no. 797
	Rosso Montecitorio, rosso Kumeta, from Monte Kumeta, Province of Palermo, Sicily, Italy	no. 85
	Giallo antico, from Chemtou, Tunisia	no. 32
	Granito, probably granito di Corsica, from Corsica, France	no. 856
	Almost certainly Proconnesian marble, marmo di Proconneso, from the island of Marmara, Balıkesir, Turkey	no. 4
	Imperial porphyry, porfido rosso antico, from Gebel Dokhan, Eastern Desert, Egypt	no. 783
80-81	Faustino Corsi from *Delle Pietre antiche: trattoto* by Faustino Corsi, 1845	Shelfmark 555.5/20
	Catalogo ragionato d'una collezione di pietre antiche by Faustino Corsi, 1825	Shelfmark 549.091/10
82-3	Samples of volcanic lava and volcanic sand from Graham Island	Daubeny Collection
84-5	*Cause and Effect* by Henry De la Beche, c.1830	William Buckland Collection
87	Letter to Mary Buckland from Elizabeth Philpot, 9 December 1833	William Buckland Collection, WB/C/P/7
88-9	Beringer Lying Stones	OUMNH T.22 and T.23
91	*Ichthyosaurus anningae*	OUMNH J.13587
92-3	*Ichthyosaurus communis*	J.13799
94-7	Various crabs collected by Charles Darwin	Bell Crab Collection

Page	Item	Collection
98–9	Thomas Sopwith's geological models, 1841	William Buckland Collection
100–101	[*Vesuvius erupting at night*] attributed to Enrico La Pira, 1845	William Buckland Collection, WB/A/D-332
102	Atlantic bluefin tuna, *Thunnus thynnus*	OUMNH.ZC.17423
104–5	Cuprite var. chalcotrichite in ferruginous quartz from Gunnislake, England	MIN.737
106–7	Trilobites from Grindrod Collection	
	Acanthopyge hirsutus (Fletcher, 1850)	C.290
	Anacaenaspis phasganis (Thomas, 1981)	C.16739
	Bumastus barriensis (Murchison, 1839)	C.17313
	Cyphoproetus depressus (Barrande, 1846)	C.783
	Dalmanites grindrodianus (Salter, 1864)	C.10
	Dalmanites myops (König, 1825)	C.35
	Deiphon barrandei (Whittard, 1934)	C.21
	Encrinurus punctatus (Wahlenberg, 1818)	C.341
	Hemiarges scutalis (Salter, 1873)	C.299
108–9	Tsetse fly, *Glossina morsitans*	DIPT0065
110	*Alfred Russel Wallace* by Emil Otto Hoppé, 1910	Museum Photograph Collection
111	Wallace's giant bee, *Chalicodoma (Megachile) pluto*	W.W. Saunders Collection
113	Letter to John Phillips from Charles Darwin, 11 November 1859	John Phillips Collection JP/C/Darwin/9
	Reproduction of portrait of John Phillips, 1866. Photographer unknown	Museum Photograph Collection
114–15	Westwood's 'giant flea'	Hope Entomological Collections
116	Ophicalcite Serpentine, County Galway	Lower North Gallery
117	Capitals and corbels	Upper North Gallery
118–19	James O'Shea at work on the cat window, 1860. Photographer unknown	Museum Photograph Collection
119	Corbel atop Carboniferous Limestone, Limerick County, Ireland	Lower North Gallery
120–21	*Mer de Glace* by Richard St John Tyrwhitt, *c.*1859	Museum Art Collection
122–3	Indian star tortoise, *Geochelone elegans*	OUMNH.ZC.08555
125	*Frederick William Hope* by Lowes Cato Dickinson, 1861	Museum Art Collection
126–7	Blaschka glass models	Blaschka Model Collection
129	*Charles Darwin* by Julia Margaret Cameron, 1868	Museum Photograph Collection
130	*The Malay Archipelago: The Land of the Orang-Utan and the Bird of Paradise: A Narrative of Travel, with Studies of Man and Nature* by Alfred Russel Wallace, 1872	203 d.23
131	*Elaphomia alcicornis*, the elk-horned deer-fly	G.H. Verrall & J.E. Collin Collection
132–3	Denny Lice Collection	H. Denny Collection
134–5	Fossil ray-finned fish, *Dapedium punctatum*	J.3001
136	Opal var. precious opal preserving a shell of the gastropod *Ampullospiro* sp.	MIN.9375
138–9	J.O. Westwood's shoes, glasses, magnifying glass and clap net	John Obadiah Westwood Collection, objects
141	Silver Bird Collection, bushwren, *Xenicus longipes* (back uppermost)	ZC.B/14539
	Silver Bird Collection, bushwren, *Xenicus longipes* (front uppermost)	ZC.B/14541
	Silver Bird Collection, bushwren, *Xenicus longipes* (lying on side)	ZC.B/14540

Page	Object	Collection
142–3	Crescent nail-tail wallaby, *Onychogalea lunata*	ZC.8020 (female), ZC.8021 (juvenile), ZC.4552 (male)
145	Aye-aye, *Daubentonia madagascariensis*	ZC.05119
146–7	Drawer of Diptera	E.K. Pearce Collection
148–9	Purple spotted swallowtail, *Graphium goodenovi*	D. Jenness Collection
150–53	Blown caterpillars	T.A. Chapman Collection
154	*Alcides aurora*	Hope Entomological Collections
155	*Papilio laglaizei*	Hope Entomological Collections
156–7	Items from the 1933 Everest expedition	Lawrence Rickard Wager Collection
158–9	West Indian Ocean coelacanth scale, *Latimeria chalumnae* (Upper)	ZC.09299
159	Part of West Indian Ocean coelacanth scale, *Latimeria chalumnae* (Lower)	ZC.09298
160–61	Magic lantern slides by J.R.R. Tolkien for Christmas lecture, 1938	Museum Lantern Slide Collection
162–3	Boxer mantis	1945-012
167	*Dorothy Hodgkin* by Anthony Stones, 2010	Museum court, Museum Art Collection
168–9	Nobel Prize Medal awarded to Nikolaas Tinbergen, 1973	Museum Object Collection
170–71	Drawer of Coleoptera	Wollaston Collection OUMNH-2006-007
172–3	Quartz var. amethyst (pale coloured sceptre crystal)	Mary Morland Collection, 29291
	Quartz var. amethyst (sceptre crystal, re-polished)	Mary Morland Collection, 29292
	Quartz var. amethyst (sceptre crystal)	Mary Morland Collection, 29293
	Axinite with quartz	Mary Morland Collection, 29344
	Schorl with apatite and quartz; stained with ferruginous clay	Mary Morland Collection, 29376
	Hematite var. specular hematite with quartz var. rock crystal	Mary Morland Collection, 29413
	Goethite var. wood goethite with quartz	Mary Morland Collection, 29418
	Rhodochrosite with alabandite	Mary Morland Collection, 29425
	Cassiterite pseudomorphous after Carlsbad twinned orthoclase (2 specimens)	Mary Morland Collection, 29431
	Malachite (acicular tufts) in bog iron ore	Mary Morland Collection, 29466
	Chalcophyllite (hexagonal crystals)	Mary Morland Collection, 29472
	Chalcophyllite in ferruginous gossan	Mary Morland Collection, 29473
	Native gold (dendritic), with quartz on a quartzose porphyry	Mary Morland Collection, 29479
	Galena	Mary Morland Collection, 29486
174–5	Catalogue of Mary Morland's mineral collection in a copy of *An Elementary Introduction to the Knowledge of Mineralogy: Comprising Some Account of the Characters and Elements of Minerals; Explanations of Terms in Common Use; Description of Minerals, with Accounts of the Places and Circumstances in which they are Found; and Especially the Localities of British Minerals* by William Phillips, 1837	Shelfmark 549/20
176–7	Otter, *Lutra lutra*	On display, Evolution Touch Table
179	Scaly-foot snail, *Chrysomallon squamiferum*	ZC 2013.02.001
180–81	Plesiosaur, *Stratesaurus taylori*	J.10337
183	Common bottlenose dolphin, *Tursiops truncatus*	ZC.19054
	Beluga whale, *Delphinapterus leucas*	ZC.21803
	Orca, *Orcinus orca*	ZC.14465
	Common minke whale, *Balaenoptera acutorostrata*	ZC.20093
184–5	Moustached kingfisher, *Actenoides bougainvillei*	ZC.B/4076
186–9	Plesiosaur, *Muraenosaurus* sp.	J.95000

Page	Item	Collection
193	Echinoderm, *Gissocrinus capillaris* (Phillips, 1839)	Museum Handling Collection, C.30628
196–7	Calotte of human, *Homo sapiens*	ZC.23167
198–9	*Encope* sp.	Charles Lyell Collection, PAL.CL.5129e
	Fusus rostratus (Olivi, 1792)	Charles Lyell Collection, PAL.CL.223f
	Cardita crassa Lamarck, 1819	Charles Lyell Collection, PAL.CL.366D
	Shark tooth, not identified to genus or species level	Charles Lyell Collection, PAL.CL.1605a
	Natica (Ampullina) sphaerica Deshayes, 1832	Charles Lyell Collection, PAL.CL.03835b
202–3	*Argina pulchra*	LEPI0355
	Blue beetle, *Buprestidae*	F.W. Hope - J.O. Westwood Collection
	Bug with pink eyes, *Membracidae*	F.W. Hope - J.O. Westwood Collection
	Papilio krishna	LEPI4019
	Thorn bug, *Membracidae*	Hope Entomological Collections
	Sarosa xanthobasis	LEPI0138
	Arcas gozmanyi	LEPI4056
	Wallace weevil, *Curculionidae*	A.R. Wallace Collection
	Red leaf beetle behind, *Chrysomelidae*	Hope Entomological Collections
	Lamprima schreibersi (green scarab)	COLE0430
	Morpho menelaus	LEPI4044
	Lantern bug, *Fulgoridae*	F.W. Hope - J.O. Westwood Collection
	White mutilid, *Mutilidae*	Hope Entomological Collections
205	*The Dodo* by Jan Savery, 1651	Museum Art Collection
207	Museum roof	Museum

The International Commission on Stratigraphy, www.stratigraphy.org, produces a useful table on geographical epochs.

Further reading

The International Commission on Zoological Nomenclature, www.iczn.org, advises the zoological community on the correct use of scientific names for animals.

Acland, H., and J. Ruskin, *The Oxford Museum*, Smith, Elder, London, 1859.

Agassiz, L., *Recherches sur les Poissons Fossiles*, I, Imprimerie de Petitpierre, Neuchatel, pp. [I]-XXXIII, 1-188, 1834.

Agassiz, L., *Recherches sur les Poissons Fossiles*, II (I), Imprimerie de Petitpierre, Neuchatel, 1835.

Anker, A., K.M. Hultgren and S. De Grave, '*Synalpheus pinkfloydi* sp. nov., a New Pistol Shrimp from the Tropical Eastern Pacific (Decapoda: Alpheidae)', *Zootaxa*, vol. 4254, no. 1, 2017, pp. 111-19.

Baerends, G.P., C. Beer and A. Manning (eds), *Function and Evolution in Behaviour: Essays in Honour of Professor Niko Tinbergen, FRS*, Clarendon Press, Oxford, 1975.

Buckland, W., 'Account of an Assemblage of Fossil Teeth and Bones of Elephants, Rhinoceros, Hippopotamus, Bear, Tiger, and Hyaena, and Sixteen Other Animals; Discovered in a Cave at Kirkdale, Yorkshire, in the Year 1821: With a Comparative View of Five Similar Caverns in Various Parts of England, and Others on the Continent', *Philosophical Transactions of the Royal Society of London*, vol. 112, 1822, pp. 171-236.

Buckland, W., 'Notice on the *Megalosaurus* or Great Lizard of Stonesfield', *Transactions of the Geological Society of London*, 2nd Series, 1, 1824.

Buckland, W., *Reliquiae diluvianae*: *or, Observations on the organic remains contained in caves, fissures, and diluvial gravel, and on other geological phenomena, attesting the action of an universal deluge*, 2nd edn, J. Murray, London, 1824.

Buckland, W., 'On the Discovery of Coprolites, or Fossil Faeces, in the Lias at Lyme Regis, and in Other Formations', *Transactions of the Royal Society of London*, series 2, vol. 3, 1829, pp. 223-36, pls 28-31.

Buckland, W., *Geology and Mineralogy Considered with Reference to Natural Theology*, 2 vols, W. Pickering, London, 1836.

Buller, W.L., *A History of the Birds of New Zealand*, John Van Voorst, London, 1873.

Carpenter, G.D.H., 'Notes by E. Burtt, B.S.c., F.R.E.S., on the Habits of a Species of *Oxypilus* (Mantidae), and the Flight of the Male of a Species of *Palophus* (Phasmidae)', *Proceedings of the Royal Entomological Society of London, Series A*, 20 (7-9), 1945, pp. 82-3.

Chandler, P.J., 'Ethel Katharine Pearce (1856-1940) and Her Contribution to Dipterology', *Dipterists Digest*, vol. 16, no. 2, 2009, pp. 117-46.

Chancellor, G., A. diMauro, R. Ingle and G. King, 'Charles Darwin's *Beagle* Collections in the Oxford University Museum', *Archives of Natural History*, vol. 15, 1988, pp. 197-231.

Chen, C., K. Linse, J.T. Copley and A.D. Rogers, 'The "Scaly-Foot Gastropod": A New Genus and Species of Hydrothermal Vent-Endemic Gastropod (*Neomphalina: Peltospiridae*) from the Indian Ocean', *Journal of Molluscan Studies*, vol. 81, issue 3, 2015, pp. 322-34.

Corsi, F., *Catalogo ragionato d'una collezione di pietre di decorazione*, Rome, 1825.
Corsi, F., *Delle pietre antiche libri quattro*, Rome, 1828.
Darwin, C., *On the Origin of Species by Means of Natural Selection: or, The Preservation of Favoured Races in the Struggle for Life*, John Murray, London, 1860.
Davies, K.C., and J. Hull, *The Zoological Collections of the Oxford University Museum: A Historical Review and General Account, with Comprehensive Donor Index to the Year 1975*, Oxford University Press, Oxford, 1976.
De Grave, S., 'A New Species of Crinoid-Associated Periclimenes from Honduras (Crustacea: Decapoda: Palaemonidae)', *Zootaxa*, vol. 3793, no. 5, 2014, pp.587–94.
D'Huarta, J.P., M.B. Nowak-Kemp and T.M. Butynski, 'A Seventeenth-Century Warthog Skull in Oxford, England', *Archives of Natural History*, vol. 40, no. 2, 2013, pp. 294–301.
Dupuis, C., 'Pierrre André Latreille (1762–1833): The Foremost Entomologist of His Time', *Annual Review of Entomology*, vol. 19, 1974, pp. 1–14.
Edmonds, J.M., and H.P. Powell, 'Beringer "Lügensteine" at Oxford', *Proceedings of the Geologists' Association*, vol. 85, 1974, pp. 549–54.
Filard, C., 'Why I Collected a Moustached Kingfisher', *Audubon*, 2015, www.audubon.org/news/why-i-collected-moustached-kingfisher (accessed 25 July 2017).
Garnham, T., *Oxford Museum: Deane and Woodward*, Phaidon Press, London, 1992.
Gordon, E.O., *The Life and Correspondence of William Buckland, D.D., F.R.S.: Sometime Dean of Westminster, Twice President of the Geological Society, and First President of the British Association*, John Murray, London, 1894.
Grierson, J., *Temperance, Therapy and Trilobites: Dr Ralph Grindrod: Victorian Pioneer*, Cora Weaver, Malvern, 2001.
Grindrod, R.B., Esq., M.D., LL.D. *The Templar; An Illustrated Temperance Treasury*, no. 132, pp. 285–9.
Lack, A.J., and R. Overall, *The Museum Swifts: The Story of the Swifts in the Tower of the Oxford University Museum of Natural History*, Oxford University Museum of Natural History, Oxford, 2002.
Lomax, D., and J. Massare, 'A New Species of Ichthyosaurus from the Lower Jurassic of West Dorset, England, U.K.', *Journal of Vertebrate Paleontology*, vol. 35, 2015.
Lyell, K.M. (ed.), *Life, Letters and Journals of Sir Charles Lyell*, 2 vols, J. Murray, London, 1881.
Marquis Di Spineto, 'On the Zimb of Bruce, as Connected with the Hieroglyphics of Egypt', *London and Edinburgh Philosophical Magazine and Journal of Science*, vol. 4, no. 11, 1834, pp. 170–78.
Morris, M., 'The Apionidae (Coleoptera) of the Canary Islands, with Particular Reference to the Contribution of T. Vernon Wollaston', *Acta Entomologica Musei Nationalis Prague*, vol. 51, no. 1, 2011, pp. 157–82.
Nowak-Kemp, M.B., '150 Years of Changing Attitudes towards Zoological Collections in a University Museum: The Case of the Thomas Bell Tortoise Collection in the Oxford University Museum', *Archives of Natural History*, vol. 36, no. 2, 2009, pp. 299–315.
Nowak-Kemp, M., and J.P. Hume, 'The Oxford Dodo. Part 1: The Museum History

of the Tradescant Dodo: Ownership, Displays and Audience', *Historical Biology*, vol. 29, no. 2, 2017, pp. 234–47.
O'Dwyer, F., *The Architecture of Deane and Woodward*, Cork University Press, Cork, 1984.
Pearce, E.K., *Typical Flies: A Photographic Atlas of Diptera Including Aphaniptera*, Cambridge University Press, Cambridge, 1915.
Phillips, J., *Memoirs of William Smith, LL.D., Author of the 'Map of the Strata of England and Wales'*, John Murray, London, 1844.
Poulton, E.B., *The Colours of Animals: Their Meaning and Use, Especially Considered in the Case of Insects*, D. Appleton and Company, New York, 1890.
Read, B., and J. Barnes, *Pre-Raphaelite Sculpture: Nature and Imagination in British Sculpture 1848–1914*, Lund Humphries, London, 1991.
Ross, A.C., *David Livingstone: Mission and Empire*, Continuum IPL, London, 2002.
Salmon, M.A., P. Marren and B. Harley, *The Aurelian Legacy: British Butterflies and their Collectors*, Harley, Colchester, 2000.
Smith, A.Z., *A History of the Hope Entomological Collections in the University Museum, Oxford with Lists of Archives and Collections*, Clarendon Press, Oxford, 1986.
Smithwick, F.M., 'Feeding Ecology of the Deep-Bodied Fish *Dapedium* (Actinopterygii, Neopterygii) from the Sinemurian of Dorset, England', *Palaeontology*, vol. 58, no. 2, 2015, pp. 293–311.
Sowerby, J., *The Mineralogy Conchology of Great Britain*, vol. II, B. Meredith, London, 1817.
Van der Meij, S.E.T., 'A New Species of Opecarcinus Kropp & Manning, 1987 (Crustacea: Brachyura: Cryptochiridae) Associated with the Stony Corals Pavona clavus (Dana, 1846) and P. bipartita Nemenzo, 1980 (Scleractinia: Agariciidae)', *Zootaxa*, vol. 3869, no. 1, 2014, pp. 44–52.
Vernon, H., and K. Ewart, *A History of the Oxford Museum*, Clarendon Press, Oxford, 1909.
Wallace, A.R., *The Malay Archipelago: The Land of the Orang-Utan and the Bird of Paradise: A Narrative of Travel, with Studies of Man and Nature*, Macmillan, London, 1869.
Weinberg, S., *A Fish Caught in Time: The Search for the Coelacanth*, Fourth Estate, London, 1999.
Weiss, M., and J. Cameron, *Julia Margaret Cameron: Photographs to Electrify You with Delight and Startle the World*, Victoria and Albert Museum, London, 2015.
Westwood, J.O., 'Observations on the destructive species of Dipterous Insects known in Africa under the names of the Tsetse, Zimb, and Tsaltsalya, and their supposed connection with the fourth plague of Egypt', *Proceedings of the Zoological Society of London*, vol. 18, 1850, pp. 258–70, pl. 19, fig. 1.
Wollaston, T.V., *Insecta Maderensia: An Account of the Insects of the Islands of the Madeiran Group*, John Van Voorst, London, 1854.

Index

Numbers of pages with illustrations are in *italics*. All institutions are in Oxford unless stated otherwise.